MW01609483

Commissioning, Preventive Maintenance, and Troubleshooting Guide for Commercial Ground-Source Heat Pump Systems

This project was supported by ASHRAE and the Geothermal Heat Pump Consortium.

Cognizant Technical Committee: TC 6.8, Geothermal Energy Utilization.

This publication was prepared by
Caneta Research Inc., Mississauga, Ontario, and Sachs & Sachs, Inc., McLean, Virginia under ASHRAE Special Project 94.

Any updates/errata to this publication will be posted on the ASHRAE Web site at www.ashrae.org/publicationupdates.

Commissioning, Preventive Maintenance, and Troubleshooting Guide for Commercial Ground-Source Heat Pump Systems

American Society of Heating, Refrigerating and Air-Conditioning Engineers, Inc.

ISBN 10: 1-931862-09-5
ISBN 13: 978-1-931862-09-7

©2002 American Society of Heating, Refrigerating
and Air-Conditioning Engineers, Inc.
1791 Tullie Circle, N.E.
Atlanta, GA 30329
www.ashrae.org

All rights reserved.

Printed in the United States of America

Cover design by Tracy Becker.

ASHRAE has compiled this publication with care, but ASHRAE has not investigated, and ASHRAE expressly disclaims any duty to investigate, any product, service, process, proce- dure, design, or the like that may be described herein. The appearance of any technical data or editorial material in this publication does not constitute endorsement, warranty, or guar- anty by ASHRAE of any product, service, process, procedure, design, or the like. ASHRAE does not warrant that the information in the publication is free of errors, and ASHRAE does not necessarily agree with any statement or opinion in this publication. The entire risk of the use of any information in this publication is assumed by the user.

No part of this book may be reproduced without permission in writing from ASHRAE, except by a reviewer who may quote brief passages or reproduce illustrations in a review with appropriate credit; nor may any part of this book be reproduced, stored in a retrieval system, or transmitted in any way or by any means—electronic, photocopying, recording, or other—without permission in writing from ASHRAE.

ASHRAE STAFF

Special Publications

Mildred Geshwiler
Editor

Erin S. Howard
Assistant Editor

Christina Helms
Assistant Editor

Michshell Phillips
Secretary

Publishing Services

Barry Kurian
Manager

Jayne Jackson
Production Assistant

Publisher

W. Stephen Comstock

Table of Contents

Acknowledgments

The development of this commissioning, preventive maintenance, and troubleshooting guide for commercial and institutional building ground-source heat pump systems would not have been possible without the financial support of the sponsors, the Geothermal Heat Pump Consortium (GHPC) and the American Society of Heating, Refrigerating and Air-Conditioning Engineers (ASHRAE).

The authors acknowledge the valuable review and comment received from members of the T.C. 6.8 Project Monitoring Subcommittee on an early version of the draft. These members include:

- David Dinse
- Gus Foster
- Steve Carlson
- Spencer Morash

The same draft was sent for external expert review. It was considered important to get this expert input on the document from experienced practitioners, to ensure the technical accuracy, relevance, and value of the resulting guide to the industry at large. Expert review was provided by:

- Robert Brown, ClimateMaster
- Robert Mancini, Mancini, Saldan & Associates
- Cedric S. Trueman, Trueman Engineering Services
- Kevin Rafferty, Geo-Heat Center

Robert Brown provided a manufacturer's perspective on the guide content. Robert Mancini provided an A/E perspective, having designed a very large number of commercial/institutional ground-source systems. Cedric S. Trueman was a member of the Guideline 1-1996 Project Committee, which drafted the *HVAC Commissioning Process* document, and provided valuable suggestions for the commissioning chapter. Kevin Rafferty reviewed and suggested revisions on sections dealing with groundwater systems throughout the guide.

The authors also wish to express their appreciation to the following companies who provided product information used in the development of Chapters 3 and 4 on maintenance and troubleshooting: Econar, WaterFurnace International, Performance Pipe, Trane, National Groundwater Association, J and P Engineers and Tracer Research Corporations.

Chapter 1
Introduction

BACKGROUND

Ground-source heat pump systems (GSHPs), also known as geothermal or geoexchange heat pump systems, are becoming more widely accepted in the commercial building sector. Existing ASHRAE design manuals (ASHRAE 1995; Kavanaugh and Rafferty 1997) have greatly advanced the state-of-the-art for design. However, they do not adequately cover technical information needed at the project implementation stage, such as special requirements for commissioning, preventive maintenance, and system troubleshooting.

As more systems enter service, scattered problems have surfaced in operating buildings. ASHRAE Technical Committee TC 6.8 has directed efforts to the production of a commissioning, preventive maintenance, and troubleshooting guide for commercial GSHP systems to serve as a companion guide to the existing design manuals. Guidance in these areas will ensure that GSHP systems deliver the energy efficiency and maintenance savings relative to competing HVAC systems that are characteristics of these systems.

COMMISSIONING

Commissioning is the process of ensuring that systems are designed, installed, functionally tested, and capable of being operated and maintained to perform in conformity with the design intent. Commissioning often doesn't get done because owners want to keep the costs down. This guide will present the commissioning requirements for GSHP systems in the design and program, construction, and acceptance phases of a commercial project.

PREVENTIVE MAINTENANCE

Preventive maintenance ensures the proper operation of a system or equipment. The objectives of a preventive maintenance program are to ensure the durability, reliability, efficiency, and safety of the system or equipment. In a new building project, functions and resources must be planned, budgeted, and scheduled. A recent investigation (Cane and Garnet 1999) concluded that a large variance of maintenance costs suggested the need for GSHP system generic maintenance guidelines. Some owners were doing too much maintenance—unnecessary work that negatively impacts GSHP cost-effectiveness. This guide presents maintenance program actions and frequency for the major GSHP system components—the heat pumps, the ground loop, the water well—which, in the absence of manufacturer information, can serve as the basis for a preventive maintenance program.

TROUBLESHOOTING

In spite of all efforts to properly install and maintain the GSHP system, equipment will break down and require fixing. Efforts over the past few years (Cane et al. 1996, 1998; Downey and Proctor 1998) have identified equipment failures, operational problems, and fixes for GSHP systems. This guide contains troubleshooting procedures for the four major subsystems making up a GSHP system. The troubleshooting section characterizes the symptoms, possible causes, and remedies or corrective actions for 32 problem areas. This information provides an excellent tool for system operators and practitioners to use in troubleshooting and fixing commercial GSHP systems.

HOW TO USE THIS GUIDE

Chapter 2 describes the process of commissioning ground-source heat pump systems in large buildings. The introduction identifies the key players and their roles and responsibilities. The chapter then details the commissioning activities in the design and program phase, the construction phase, and the acceptance phase. Sidebars or tables are used throughout Chapter 2 to serve as reminders of items related to commercial ground-source heat pump design, prefunction checks, and functional testing around which a commissioning specification can be written.

Chapter 3 details preventive maintenance requirements for large ground-source systems. Maintenance frequency and actions are provided for heat pump units, fan-coil units (for central systems), and ground-loop and groundwater system components. These have been derived from manufacturer recommendations but are not intended to replace specific manufacturer requirements where available.

Chapter 4 presents a troubleshooting guide for large ground-source heat pump systems. A total of 34 symptoms, with an even greater number of possible causes, are presented for the reader. In the case of each possible cause, there is a proposed remedy. A numbering system is included to guide the reader between these tables and an appendix that provides detail on selected proposed remedies.

CONTEXT

In Chapter 2, we maintain that the architect/design engineer (A/E) can act as the commissioning authority on GSHP systems, particularly in smaller buildings, because of their inherent simplicity. In larger buildings, this is only realistic if the A/E has detailed system knowledge, experience in planning and documenting test procedures, extensive field experience, and good communication skills. If this is not feasible a separate commissioning authority should be hired by the owner. Throughout the text the use of "commissioning authority" may be interchanged with A/E if the intent is that the A/E will act as the commissioning authority.

Chapter 2
Commissioning

Commissioning is the process of ensuring that systems are designed, installed, functionally tested, and capable of being operated and maintained to perform in conformity with the design intent. In this context, the primary goal of this chapter is to document the necessary steps to ensure that the design intent is appropriate, understood, and fully implemented and that the implementation can be verified to the satisfaction of the owner.

The chapter is structured as follows.

- Introduction: commissioning and the GSHP system
- Design and program phase activities
- Construction phase activities
- Acceptance and post-acceptance phase activities
- Conclusion

Throughout this chapter, the term "A/E" is used to refer to the design team, including the architect and mechanical/electrical engineers.

INTRODUCTION TO THE COMMISSIONING PROCESS

For conventional systems, commissioning is often not done because owners and developers want to keep first costs down. As systems have become more complex over time, fully functional operation is unlikely without implementing an organized commissioning process. Thus, commissioning has become important in achieving some important objectives:

1. Ensure that the owner's intent is understood and translated into a buildable, testable, and maintainable system.
2. Provide a way to ensure that indoor air quality, efficiency, building code, and similar issues are fully reflected in the design and construction of the structure.
3. Enable a quality-focused approach to be implemented.
4. Clarify, from a legal perspective, the roles and responsibilities of each participant in the construction and installation process.

The Commissioning Process

There are several distinctions between commissioning and customary test-and-balance approaches. Commissioning is a process rather than a discrete activity. It focuses on the system as well as the individual components, and the focus is to deliver both verified performance and accountability. The process of commissioning as developed in this chapter is presented in Table 2-1.

Table 2-1
Summary of Commissioning Process[a]

Function	Performed by	Reviewed by	Witnessed by
Program and Design Phase			
Document Owner's Program, Design Intent, and Basis of Design	A/E		
• Develop the Commissioning Plan	Commissioning authority		
• Develop the Commissioning Specification	Commissioning authority		
• GSHP - Specific commissioning topics	A/E	Commissioning authority	
Construction Phase Activities			
• Synchronize Commissioning Plan with design changes	Commissioning authority		
• Construction, installation, and prefunctional testing	Contractor/TAB	A/E, commissioning authority	
• Startup and initial operation	Contractor/manufacturer		Commissioning authority
Acceptance Phase Activities			
• Functional testing	Contractor/manufacturer	A/E	Commissioning authority
• Verification	Commissioning authority	A/E	Commissioning authority
• Complete the commissioning report	Commissioning authority	A/E	
• Complete the systems manual and record documentation	Contractor	A/E, commissioning authority	A/E, commissioning authority
• Train the operation and maintenance staff	Contractor/manufacturer		Commissioning authority

a. **Principal Commissioning Sources.** Principal documents in the commissioning area are ASHRAE (1996) and PECI (1998). These two documents are designed for large complex buildings. Thomas and Madgett (2000), a detailed guide specification for U.S. federal ground-source projects, gives generic language for many parts of projects but has little on commissioning per se.

This chapter distills the essence of these documents, tailored to the relative simplicity of ground-source heat pump systems. Thus, it does not include all of the steps covered in these documents but provides guidance appropriate to GSHP commissioning.

Note: "A/E" and "commissioning authority" are separate entities in Table 2-1. Where the building is relatively small or the A/E is qualified, willing, and compensated, the commissioning authority can be the A/E. Table entries would change appropriately.

Proponents maintain that using a commissioning process has significant legal advantages.[1] A carefully drawn commissioning specification can clarify the engineer's responsibilities and legal liabilities. This includes legal exposure at the work site with respect to safety and other issues. It has been shown that disciplined commissioning can save money in the long term and avoid major problems that can lead to legal claims.

Consequently, a small industry has developed that advocates and does commissioning. Proponents generally recommend engaging an independent "Commissioning Authority" for large commercial/institutional buildings with complex equipment and control systems, particularly for government owners.

This guide advocates an alternative approach for buildings served by GSHP systems, an approach based on the inherent simplicity of GSHP systems. We maintain that commissioning lies within the traditional responsibility of the design engineer when a disciplined design and construction program, focused on simplicity, is applied to GSHP systems, particularly in small buildings. A separate commissioning authority should not be needed for these GSHP systems. The key is to minimize complexity and seek simplicity for performance, control, efficiency, and maintainability. However, this approach in larger buildings is only realistic if the following conditions apply:

- The designer's personnel must have detailed system knowledge, have experience in planning and documenting test procedures, have extensive field experience (the on-site person must be comfortable being in the field), and have good communication skills.
- The commissioning responsibilities need to be included in the designer's terms of reference and be compensated for.

If this is not feasible, a separate commissioning authority should be engaged by the owner and the roles and responsibilities of the various players would be as depicted in Table 2-1. Otherwise, the commissioning authority would be replaced by the A/E in Table 2-1.

GSHP System Components

In this discussion, the following system architecture is assumed.

Ground System. The ground system can be either a horizontal ground heat exchanger, vertical ground heat exchanger, or a groundwater well system. The purpose of the ground system is to reject or absorb heat to or from the ground or groundwater. The components of the ground system include the ground heat exchanger or wells, run-outs, headers, manifolds, flow control valves, pumps, strainers, and plate-frame heat exchanger (for well systems).

Distribution System. The distribution systems includes the piping that distributes the ground energy to the heat pumps and the ductwork that distributes the

---

1. Hornreich (1994); Tyler (1994).

warm or cold air from the heat pumps to the conditioned space. The distribution system is largely piping with zone ducts when small packaged heat pumps are distributed throughout the building and piping and fan-coils when central heat pumps are employed. These systems are known as "distributed" and "central station," respectively. The distribution system, therefore, includes piping, pumps, expansion tank, air vents, ductwork, supply diffusers, return grilles, flow control dampers, and fire dampers.

Heat Pump Unit. The heat pump unit consists of the heat pump, with all its ancillary equipment, such as hose kits, valves and controls, thermostat, electrical disconnect, etc.

Ventilation System. This is the system that supplies outside air to the building. Typically, the ventilation system consists of an outside air intake, heat pump or other conditioning means and energy recovery ventilator (ERV), ductwork, control dampers, fire dampers, and controls. Since an exhaust air system is required for the ventilation system to supply adequate air, it is really an integral part of the ventilation system, even if no ERV is fitted. Exhaust air systems are made up of exhaust grilles, ductwork, fans, and dampers. In many situations, separating ventilation air and its preconditioning from control of conditions within zones simplifies systems.[2]

Table 2-2 presents an example ground-source specific commissioning process showing functions, roles, and responsibilities of the various players for the case where the commissioning authority and A/E are separate entities. Where the A/E is qualified, willing, and compensated, the commissioning authority would be the A/E in Table 2-2 and in the text that follows in this chapter.

PROGRAM AND DESIGN PHASE ACTIVITIES

Four commissioning tasks prior to construction are necessary. These are (party responsible):

1. Gain agreement on the owner's objectives in terms of physical parameters (temperature, relative humidity in spaces), occupancy needs, energy consumption, etc. Document the design intent accordingly, and formalize it further in a document giving specific technical criteria (A/E).

2. Begin developing the Commissioning Plan as a verification instrument (commissioning authority).

3. Develop the Commissioning Specification (commissioning authority).

4. Check that the *design review* process accommodates commissioning. This includes verifying that the design will achieve the design intent, that the schedule allows commissioning, and that the documents include the commissioning process (commissioning authority).

---

2. Coad (1996).

Table 2-2
Example Ground-Source Commissioning Process for Mechanical Design

System	Function	Performed by	Witnessed by
Heat Pump Water Piping	Pressure test, clean, fill and purge air	Contractor	A/E
Ground-Source Piping	Pressure test, clean	Contractor	A/E
	Fill and purge air	Contractor	—
Pumps	Inspect, test, startup	Contractor	—
Heat Recovery Unit	Inspect, test, startup	Manufacturer	Commissioning Authority
	Provide clean set of filters	Contractor	—
	Staff instruction	Manufacturer	Commissioning Authority and Owner Representative
Heat Pump Units	Inspect, test, startup	Manufacturer	—
	Provide clean set of filters	Contractor	—
	Staff instruction	Manufacturer	Commissioning Authority and Owner Representative
Chemical Treatment	Flushing and cleaning	Contractor	A/E and Commissioning Authority
	Chemical treatment	Contractor/manufacturer	—
	Staff instruction	Manufacturer	Commissioning Authority and Owner Representative
Air and Water Balancing	Balancing	TAB Contractor	—
	Spot checking	TAB Contractor	A/E and Commissioning Authority
	Follow-up site visits	TAB Contractor	Commissioning Authority
Controls	Contractor installation/commissioning	Contractor	—
	Staff instruction	Commissioning Authority	Commissioning Authority and Owner Representative
	Performance testing	Commissioning Authority	—
	Seasonal testing	Commissioning Authority	—

Note: A/E and commissioning authority are separate entities in Table 2-2. Where the building is relatively small or the A/E is qualified, willing, and compensated, the commissioning authority can be the A/E. Table entries would change appropriately.

Document the Owner's Program, Design Intent, and Basis of Design

ASHRAE (1996) subdivides the beginning of the commissioning process into three initial stages:
1. Owner's Program
2. Design Intent
3. Basis of Design

The Owner's Program lays down the intended uses of the building and the performance expectations such as energy consumption and comfort specifications. The Design Intent is a more detailed description of the Owner's Program. ASHRAE (1996) gives a suggested table of contents for a Design Intent document. From this table of contents, the items applicable to ground-source systems are:
1. Functional use of facility
2. Occupancy requirements
3. Quality of materials and construction
4. Environmental and air quality requirements
5. Energy performance criteria
6. Description of all operating systems
7. Statement of each system's operation under normal occupancy, partial occupancy, emergency situations, etc.
8. Acceptable performance criteria and operating strategy for each system
9. Defined limits and restrictions for the facility
10. Warranty issues
11. Budget considerations and limitations

The Basis of Design is the formal list of all the technical criteria required to meet the Design Intent. The owner should indicate an interest, preference, or requirement for a GSHP system within either of these documents. Alternatively, the designer may have undertaken the investigation as part of evaluating HVAC options. In either case, the interest, preference, or requirement should be included in the Design Intent or the Basis of Design.

Begin Development of the Commissioning Plan as the Verification Instrument

The Commissioning Plan is the overall plan, developed before or after bidding, that provides the structure, schedule, and coordination planning for the commissioning process. The Commissioning Plan is an evolving document, outlined in the earliest phases of the project and given greater detail as the design becomes more specific and construction proceeds. In the earliest stages, the Commissioning Plan is conceptual, focusing on what is required to commission the systems being considered and who is to be involved in commissioning. One goal of formalizing the process is to heighten awareness that the owner and the owner's facility management and maintenance personnel have an important role to play. This role is in approving systems, ensuring that the plans for their placement and installation allow maintenance, and that the need for proper training and documentation requirements are included in the Commissioning Plan.

ASHRAE (1996) includes a list of 18 components of the design-phase Commissioning Plan. They range from developing the detailed requirements for the Commissioning Specification (CSI Specification 17100) to sample document and manual formats.

Develop the Commissioning Specification

The Commissioning Specification is the document that stipulates what testing will be done and by whom, when, and with what supervision. A sample Commissioning Specification is given in ASHRAE (1996), and a more detailed version is given in PECI (1998) in the context of commissioning the HVAC system (Division 17100). Appendix A-1 of this publication is a simplified Division 17100 guide for ground-source heat pump systems. However, additional commissioning requirements are also found in other divisions. These include Division 2 (site work), Division 3 (borehole grout for GSHP systems), Division 15 (HVAC systems), and Division 16 (electrical).

Contractor responsibilities related to commissioning will include

- incorporating commissioning activities into the overall construction schedule and informing the commissioning agent;
- carrying out prefunctional tests and equipment startup in accordance with checklists or procedures in the Commissioning Plan, and included in the specifications as appropriate;
- Coordinating schedule with the commissioning agent;
- operating equipment and systems under the direction of the commissioning authority so that the commissioning authority can witness the functional performance tests and verify operation and performance;
- being responsible for testing, adjusting, and balancing work for hydronic and air distribution systems (subcontract to specialist balancing agency).

The commissioning-related work to be done by the contractor will be included in all the sections and divisions of the specification that relate to the systems being commissioned.

Appendix A-2, from the table of contents of Thomas and Madgett (2000), suggests the large range of topics that might require commissioning attention for complex systems. It is expected that relatively few of these areas will be included in typical GSHP design documents. Even as early as the design phase of the project, the Commissioning Plan and the Commissioning Specification should note the required construction checks and the variety of acceptance tests that may be required.

GSHP-Specific Commissioning Topics

Requirements for commissioning the GSHP-specific components (heat pumps, water loop(s), and ground system) are outlined in CSI Division 15, Section 15995. Prefunctional tests are in Section 15997 and functional tests in Section 15998. Critical issues for GSHP systems include:

Interior Details. This includes piping layout (topology and sizes), pumps and controls, air vents, strainers, expansion tanks, makeup lines (and meters), etc.[3] Sidebar 2a lists some issues and resources.

Heat Pumps. In general, ground-source heat pumps should be extended range (formerly called ARI-330, now ISO 13256-1 Ground Loop Heat Pumps). "ARI-320" (ISO 13256-1 Water-Loop Heat Pumps) are rated under a very narrow set of liquid temperatures (86°F standard cooling, 68°F standard heating) and are not considered appropriate for ground-source applications.

GSHP units should be high efficiency with a recommended minimum EER of 14. On an annual energy basis, most commercial systems reject more heat to

---

Sidebar 2a: Commissioning Issues for Interior Details

- Review pipe sizing and layout to ensure low pressure drops and adequate flow for all heat pumps. Make sure branch lines and hose kits are adequately sized.
- Minimize air vents in pipe systems. Avoid piping configurations that require a vent at each heat pump. Use manual rather than automatic vents.
- Strainers central (in mechanical room) or distributed (at each heat pump). Consider maintenance implications of strainers at plenum-mounted heat pumps.
- Expansion tank (if any) properly located and adequately sized.
- Meter and appropriate backflow prevention valve on makeup water line, if so equipped.
- Circulating pump, variable-frequency drive (if applicable), and controls.
- Chemical treatment of closed-circuit fluids (check properties).
- Antifreeze concentration.
- Adequate flow balancing devices (circuit balancers).
- Adequate slope in condensate drains.
- Integrity of pipe and duct insulation.
- Interference with other services.
- Check that proper service space is provided around equipment.
- Check that all heat pump safety devices are installed.

---

3. ASHRAE (1995); Kavanaugh and Rafferty (1997).

Sidebar 2b: Design Considerations

Evaluating Ground System Design Options. ASHRAE (1995, chapter 2) discusses the importance of a site survey in selecting among options. Sachs (2002, Chapter 3) provides a decision tree to help select among options by site-specific feasibility and projected cost. This discussion focuses on design-phase implications of common options, the ground-coupled (closed-loop) vertical borefield, surface water (pond loop) systems, and groundwater (open loop) systems.

- **Borefield.** Establish the test borehole requirements prior to design. Design-phase issues include assessing the adequacy of the thermal modeling of the borefield[a] and ensuring that drilling, trenching, and pipe laying complies with state and local regulations. Grouting, backfilling, and erosion control are also subject to regulation.

Some Useful References for Borefield Design Issues

- Pipe specification: (IGSHPA 1997) and CSA C448.
- Grouting: (IGSHPA 1991, 2000; EPRI 1997)
- State and local regulations: McCray (1997) provides guidelines that may be adopted in some states and localities. Den Braven maintains a web site with state and local regulations as of 1999.
- Geological and drilling issues: Sachs (2002), Rafferty (Rafferty 2000)
- Field design issues for commercial-scale systems: ASHRAE (1995) and Kavanaugh and Rafferty (1997).
- Related specifications: Thomas and Madgett (2000) for backfill (Section 02200), trenching (Section 02225), and erosion control (Section 02270). Division 3 (Section 03600) deals with thermally enhanced bentonite grout but does not include thermally enhanced (or other) cement grout.

Selection of type of ground exchanger begins with site evaluation and testing. In some instances it may not be feasible to use certain configurations.

- **Surface Water System Design.**[b] Design topics include verifying adequate size of the surface water body (and/or water flow) for the load, designing for proper head loss, and ensuring that the heat exchanger can be installed and will be adequately protected from damage from surface traffic and water flow.

Sidebar 2b: Design Considerations (Continued)

- **Groundwater System Design.**[c] There needs to be a bid specification for the well and its development. The water supply (and
injection) well design is then based on the production capability
established during well testing. The design specifications should
require production of water of (chemical) quality comparable to
that produced by the test well. The specifications also should
require that the well water meet a level of quality with respect to
lack of sand production.[d] Water should not be passed through
heat pump heat exchangers but through a plate frame heat
exchanger first. Isolate from the building loop. Employ a
hydrogeologist if unfamiliar with water wells.

  a. Design review should include steps to ensure that the field models have been run
     for long enough (10 years minimum) to show that the field temperature remains
     acceptable.
  b. ASHRAE (1995, chapter 11) and Kavanaugh and Rafferty (1997, chapter 7).
  c. Thomas and Madgett (2000) offer specifications related to wells (Sections 02520
     and 02525).
  d. The specifications should require "sand-free" water production (< 1 ppm for sys-
     tems using injection wells, < 10 ppm for systems with alternative disposal wells,
     according to Kavanaugh and Rafferty [1997], p. 90).

---

the ground than they extract. In addition, the heat of compression of the machinery is rejected to the ground. The greater the cooling efficiency of the equipment,
the less heat rejected and the lower the required ground system heat flow capacity
(including a smaller ground heat exchanger). The smaller ground system and heat
exchanger will reduce first costs, even though the higher efficiency heat pumps
are more costly. "High efficiency" does not just mean a high EER and COP but
also low pumping losses for the heat pump circuit.

The A/E should ensure that installation issues are addressed. These include
equipment locations that allow easy access for service and maintenance, which
will reduce maintenance costs and servicing difficulties.

Consider requiring submittals that include the heat pumps and key auxiliary
components being provided by one manufacturer. This would include the two-
way valves (if a variable frequency drive/pump is used), flexible hose kits for
connecting the units, shutoff valves, balancing valves if used, and zone thermostats. Single-sourcing minimizes the likelihood of finger-pointing in cases where
it is suspected that, for example, electrical spikes from two-way valves might be
affecting control wiring in the heat pumps.

Subcontractor Interfaces. In closed-loop (ground-coupled and surface
water) systems, it is common that the loop installer is not a subcontractor to the
mechanical contractor. Where these entities report separately to the general contractor, it is imperative to carefully delineate the physical interface (typically the
manifolds in the pump room) and the responsibilities for flushing and purging

both the field (borehole or surface water heat exchanger) and the building loops. In design review, the A/E must check that these responsibilities are carefully defined. Analogous questions arise in terms of connecting the production (and discharge) well systems to the mechanical system. In this case, the A/E should specify that the mechanical contractor is responsible for the installation of the lines from the wells and plate-and-frame (or other) heat exchanger that isolates the groundwater from the building loop. Independent verification of the cleaning and flushing of the systems is recommended to avoid arguments when problems occur.

A Final, Key, Step. In the context of verification, arguably the most critical part of developing the Commissioning Plan and the role of the commissioning agent is *design review*—the process of reviewing and accepting contract documents, before they go out to bid, for compliance with the Design Intent documented above.

CONSTRUCTION PHASE ACTIVITIES[4]

There are three areas of primary focus for ground-source heat pump system construction-phase commissioning activities:

- Keep the commissioning plan synchronized with design changes that inevitably occur.
- Observe construction, installation, and prefunctional testing (i.e., startup, operation, and testing, adjusting, and balancing [TAB]).
- Oversee operations and maintenance (O&M) training.

Synchronize Commissioning Plan with Design Changes

It has been asserted that no building has ever been constructed according to its design and plans. Somehow, there are inevitable changes that lead the building as constructed to deviate from its drawings. Further, no set of "as-built" drawings ever completely reflects the actual installation, because the pressures on the trades simply preclude full documentation. These deviations are taken as a matter of course in commercial construction. Thus, the first commissioning activity in the construction phase is review of submittals and careful inspection to ensure that the equipment received and installed matches the contract and submittals. This is the design engineer's responsibility as well. Most changes occur due to site conditions rather than lack of review by the engineer.

Commissioning also aims to facilitate complete and accurate records by putting more emphasis on documenting changes as they happen and by involving the subcontractors as fully as possible in the process of ensuring the integrity of the as-built drawings. Another goal is to ensure the owner's O&M personnel gain the best possible understanding of the systems they will be operating and main-

---

4. ASHRAE (1996, Section 7); PECI (1998, Division 17100, particularly part 3.4).

taining. This is particularly important with systems, such as GSHP, they may not have seen before. Toward this end, the owner's O&M personnel should receive information on the basics of the overall system and its major components, observe and monitor installation of the HVAC systems, and be able to ask questions and have them answered during this time. As important, they may observe aspects of installation that will make maintenance or access difficult or impossible.

Because changes in equipment will often require changes in the schedules for testing (prefunctional checks, functional tests, etc.), it is important that the Commissioning Plan be continually revised to reflect new schedule requirements.

Observe Construction, Installation, and Prefunctional Testing

In the commissioning process, observations of construction and installation have three important goals:

- Quality control
- Observe startup, operation, and test, adjust, and balance (TAB)
- O&M program development

Quality control requires making sure that the right equipment is installed and that it is installed according to the manufacturer's directions, applicable codes, and standards. This requires verifying equipment against submittals and plans and updating the plans when substitutions or changes are made. It also requires witnessing activities pertinent to GSHP systems, such as flushing and purging water loops of debris, dirt, and air. Also, some components require prefunctional checks or tests prior to, or in conjunction with, startup. Sidebars 2c to 2i list items that should be checked for conformity to the design, installation, and prefunction test specifications and standards.

Sidebar 2c: Horizontal Ground Heat Exchanger Prefunction Checklist

Check that the following meets the construction and installation specifications and standards:

- Storage of piping and fittings on-site
- Excavation of trenches
- Bedding of trenches
- Layout of pipe in the trenches
- Fusion joints of all HDPE pipe sections
- Fusion joints to headers
- Tracer tape layout
- Backfilling, layering, and compacting of trenches
- Flushing and purging
- Check loop as-built site map
- Layout of loop field

Sidebar 2d: Vertical Ground Heat Exchanger Prefunction Checklist

Check that the following meets the construction and installation specifications and standards:

- Drilling of boreholes
- Pre-filling and capping of U-tube heat exchangers
- Tremie-grouting of U-tubes
- U-tube hydrostatic test procedure and results
- Excavation of run-out trenches
- Placement of run-outs in trenches

- Fusion joints to run-outs and headers[a]
- Circuits are within allowable length differences
- Run-outs are within allowable length differences
- Run-out pressure test procedure and results
- Disposal of water and sediment used in borehole development
- Backfilling and compacting of trenches
- Flushing and purging
- Check depth of boreholes and pipe installed
- Check for obstructions in vertical piping
- Verify strength of pipe used
- Check for loop as-built site map
- Layout of loop field

a. In general, International Ground-Source Heat Pump Association (IGSHPA) requires heat fusion for all joints. However, as of this writing, two manufacturers have applied for certification of mechanical "stab" fittings adapted from the natural gas industry.

Prefunctional checks are required for items that need to be physically complete for safe startup and for tests of controls or other components to ensure safe and controllable operation after startup. Often, many different trades are involved in achieving startup readiness. A well-organized prefunctional check/test process, carried out in cooperation with a competent and quality-conscious contractor, will minimize delays and on-site inefficiencies caused by items of work not being completed properly by the agreed schedule.

For most HVAC technologies, particularly close attention to control systems (and building automation systems) is necessary. Well-designed GSHP systems operate with minimal controls.

One might consider a distributed two-pipe heat pump system with two-way valves on the heat pumps and a variable-frequency drive on the circulating pump with differential pressure to control the pump and zone thermostats to control the individual heat pumps. Monitoring points may be installed to allow monitoring, but there need not be controllers unless scheduling or demand limiting is required. Scheduling of heat pumps, circulating pumps, and unit fans may be appropriate. For open-loop systems, well pump control is particularly important.

Sidebar 2e: Groundwater System Prefunction Checklist[a]

Check that the following meets the construction and installation specifications and standards:

- Location,[b] depth, and diameter of well(s)
- Well drilling methods
- Pump
- Construction details[b]
  - Casing
  - Screen(s)
  - Gravel pack
  - Formation seal
  - Pitless adapter
- Well development methods[c]
- Treatment/disinfection (if required)
- Well yield[c]
- Water quality
- Check for contaminants in the soil
- Well completion and flow test reports

a. These items are subject to modification on the basis of downhole conditions discovered in the drilling program.
b. Relative to other site features and service items, such as distance from foundations, water bodies, and septic systems.
c. Any change in efficiency will require recalculating the size of the ground heat exchanger and related components; in commercial systems, reduced efficiency is likely to require a significantly larger borehole field, well pump, or equivalent.

Sidebar 2f: Heat Pump Unit Prefunction Checklist

Check that the following meets the construction and installation specifications and standards:

- Unit type, capacity, and efficiency[a]
- Hangers, vibration absorbers or pads
- Balancing/two-way valves
- Piping run-outs to unit
- Hose kits and shutoff (isolation) valves
- Condensate collection and disposal system
- Position of the heat pump unit with respect to noise concerns
- Timing of connection to main loop (after flushing and purging)
- Electrical connections and disconnects
- Thermostat and other controls
- Restrictions for temporary heating or cooling during construction
- Filter removal is possible
- Heat pump access
- Heat pump removal space
- Removal of all compressor, blower, or packaging supports
- Verification of pressure/temperature (p/t) ports for servicing
- Check condensate pan and piping pitch

a. Any change in efficiency will require recalculating the size of the ground heat exchanger and related components; in commercial systems, reduced efficiency is likely to require a significantly larger borehole field, well pump, or equivalent.

Sidebar 2g: Ductwork Prefunction Checklist

Check that the following meets the construction and installation specifications and standards:

- Location of supply, return, and exhaust ducts
- Location and type of supply diffusers, return grilles, and exhaust grilles
- Duct hangers
- Flexible connections
- Duct liners
- Dampers
- Fire dampers
- Outside air connections to heat pump
- At least 90° turn in supply to alleviate noise problems
- Check that the air filter is accessible

Sidebar 2h: Piping Prefunction Checklist

Check that the following meet the construction and installation specifications and standards:

- Location of pipes
- Size of pipes
- Thermal expansion allowed for
- Cleaning and protection from construction debris (strainers)
- Hangers and isolators
- Wall sleeves
- Air vents, drain cocks, valves, and strainers
- Thread sealants
- Copper solder joints
- Insulation for loops expected to operate at < 12°C
- Isolation valve type and location
- Two-way valves on heat pumps where specified
- Balancing valve type and location
- Expansion tank
- Check for adequate piping insulation on loops expected to operate at < 55°C

Sidebar 2i: Pumping Station and Mechanical Room Equipment Prefunction Checklist

Check that the following meets the construction and installation specifications and standards:

- Pump is connected to mechanical room headers according to design
- Alignment done
- Pump and motor lubricated as required
- Pump rotates freely
- Disconnect switch installed
- Correct fusing and thermal overload protection installed
- Electrical connections complete
- Type of reducers and increasers
- Minimum radius of elbows
- Support of suction and discharge piping
- Certified pump performance test
- VFD operation
- Redundant pump automated switch-over (if installed)
- Location of expansion tank and makeup water line and alarm
- Location of air vents, blowdown drain valves, strainers
- Location of the connection to the building water loop
- Check valves
- Pump shutoff/isolation valves
- Headers
- Header service ports
- Run-out balancing valves
- Pressure safety valves (if required)
- Expansion joints
- Strainers
- Vibration isolation base and spring hangers
- Pressure test procedure and results
- Run-out isolation valves
- Antifreeze type, density, and inhibitors with labels, as required
- Flushing connections
- Balancing valves
- Check that air separator is downstream from building loop to trap air from unit maintenance activities before it can flow to earth loop
- Check antifreeze type and density for correct freeze protection if needed
- Verify open-loop heat exchanger connections are in accordance with design
- Verify open-loop screen hole size in groundwater strainer

Observe Startup, Operation, and <u>Test, Adjust, and Balance (TAB)</u>. The GSHP system startup should be undertaken only after all prefunctional tests (see Appendix C) are completed and problems remedied. An additional pre-startup inspection and checkout is essential to ensure that all heat pumps, other equipment, and components are in accord with design specification, including model number, voltage, amps, etc., that they are installed properly, and that they are clean, adjusted, and ready to operate. Sidebar 2j lists an example startup procedure for a GSHP.

O&M Training Program Development. As noted above, it is highly recommended that the owner's O&M personnel participate in the observation activities, as part of their training to operate the systems. They also bring unique insights to bear on the construction process. Actual training may begin in the construction phase or during acceptance testing. The goal of the training program is to give the building operators the knowledge required to operate and maintain the systems so they achieve the design intent. This requires:

1. Understanding the HVAC systems, particularly the ground system portion of it.
2. Operating procedures for the system and all its equipment.
3. Theory of operation of system, including key temperatures and flows.
4. Maintenance (scheduled and unscheduled) for the system and all its equipment.
5. System and equipment diagnostics.
6. Reports and records.

Although formal training in the classroom and the building is useful for complex buildings, for relatively simple GSHP systems brief training and a complete manual should suffice, especially when combined with training videos of the operating sequence of each piece of equipment. The major responsibilities of the A/E in this phase of commissioning are to ensure that the manual is complete, designed for usability, and that arrangements are made for appropriate training by manufacturers and others during the acceptance phase.

Construction Phase Summary

Observation during the construction phase of a GSHP system is basically the work that the commissioning authority would want to do to ensure that the building will operate according to the design intent. It includes witnessing prefunctional and functional testing and spot-checks of reported TAB results for air and water systems, as well as the quality-oriented observations emphasized above. Construction phase commissioning activities are not complete until this work is done. This work adds value beyond the bare-bones specification and planning work to which designers are sometimes confined, and should be compensated accordingly.

ACCEPTANCE PHASE ACTIVITIES

The transition from construction phase commissioning activities to acceptance phase activities occurs after the systems are up and running. Functional testing can start after the system is believed to be operating in agreement with the contract documents.

Sidebar 2j: Example Startup Procedure for a GSHP System

1. Before starting the pumping station
    *   ensure pump rotation is correct and
    *   pump status indicators, both local and remote, are working.
2. Start the pump and ensure that system pressures are correct, that vibration and noise are acceptable. Record motor volts and amps: rated and measured, by phase, if applicable.
3. Establish the total system flow and ensure that the water-loop temperature is within the limits of the design. Do not operate the heat pumps during this period.
4. Verify proper voltage to all heat pumps and other equipment.
5. Start with the heat pump unit closest to the central pumping station. Close that unit's disconnect with the heat pump turned off at the thermostat or system controller. Start the unit fan by selecting fan "on." If the fan does not start, do not attempt to start the heat pump until troubleshooting has identified the problem.
6. Turn on the heat pump in cooling or heating mode to verify output at the heat pump heat exchanger by measuring pressure and temperature differentials and using manufacturer specifications to confirm output within specification; output should be compared with manufacturer's ratings for the same liquid temperature, inlet air temperature, and airflow rate, for both heating and cooling cycle.
7. Try opposite cycle and check output against manufacturer's specification, as before.
8. Repeat this process for all heat pumps on the water-loop.
9. Where units are not performing as called for by manufacturer specifications, determine the reasons and take appropriate steps to rectify the problems.
10. On completion, ensure that all thermostats, system controllers, and equipment are set to the design settings, and check operation of safety devices.
11. Liquid temperature difference across each heat pump heat exchanger shall be within the range specified in the design.
12. Ensure all unit valves are open.

Other acceptance phase activities can be characterized as "tying up loose ends." They include verification (ensuring that the system works and complies with the contract document and that the TAB report is accurate); completion (commissioning report, systems manual, as-built records); and turning over the building to the owner.

Training the building operators is very important. It is included in the acceptance phase since that is when the bulk of the activity takes place.

Functional Performance Testing

Functional testing ensures that each subsystem is performing according to the Design Intent, and that the startup process from system completion to full operation is conducted in a logical, systematic fashion. The cause of performance deficiencies can then be more easily identified and corrected before the building is in service. Functional testing should be done for all modes of operation—full load, part load, power failure, etc. Some functional performance tests for GSHPs are shown in Sidebar 2k.

Sidebar 2k: Some Functional Tests for GSHP Systems

1. Control system operation in all modes.
2. Power consumption, voltage, amps for all heat pumps, all circulating pumps, heat rejector fans and pumps (if used), well pumps (if used), and ventilation fans.
3. Liquid or water temperatures and flow rates not checked and recorded during balancing.
4. Airflow rates and temperatures not checked and recorded during balancing.
5. Spot checks of liquid and airflow rates and temperatures checked and recorded during balancing.
6. Operation of the control valve used to bypass (if used) the ground heat exchanger, well-water plate heat exchanger, or surface water heat exchanger when the building is in thermal balance.
7. Performance of plate-frame or other heat exchanger that isolates building loop from ground water. Include head loss (both circuits) and effectiveness (from approach temperatures).
8. Capacity and efficiency checks on all heat pumps installed on the water-loop.
9. Capacity checks on the heat rejector (if used).
10. Check that the proper concentration of antifreeze and/or water treatment chemicals are in the water loop.
11. Check of startup and shutdown procedures and sequences.

Depending on the design, this should be relatively simple for GSHP systems. However, some alternatives will impose some special requirements for GSHP systems. The best example is the need to test and verify performance of wells and pump systems for groundwater heat pump systems. This includes permits, production capability (done as part of well installation), rejection capability (if an injection well is used), water quality, and sand production (ppm), etc.

As for almost all other system types, the A/E and the construction team need to consider how to handle seasonal testing for equipment and system conditions that can only be verified during specific seasons. In some cases, part-load tests can be used, with extrapolation based on manufacturers' data for the equipment, but this may not be adequate in all cases (as an example, in some cases it may be difficult to fully check out the ventilation systems). This is less likely to be troublesome for GSHP systems that are based on unitary equipment. It is infrequent to find variable output (or two-speed) heat pumps in commercial applications, so testing only requires verifying heating and cooling peak performance. For the water circulating system, operation of the VFD and pump requires measuring maximum and minimum flow and power requirements. Procedures for the ventilation systems will vary according to design.

Once the functional tests are completed, test, adjust, and balance (TAB) procedures can begin. With the control system operation verified in the functional tests, TAB can begin with the confidence that each component in the system is operating correctly. Only then can the system be properly balanced. There are no special TAB requirements for GSHP systems, and it is often simpler than competing HVAC systems. TAB should be performed under direct supervision of the commissioning agent and should be contracted directly by the owner.

Verification

The primary tasks of verification are to review the tests that have been done (TAB, automatic control system tests, etc.) and carefully evaluate that everything, including the controls, works under all foreseeable operating conditions.[5] A major goal of this work is to ensure that the GSHP system complies with the contract document. These activities should be relatively simple for GSHP systems.

Complete the Commissioning Report

The final commissioning report[6] will document changes between the contract and the final as-built product. It should include:

- The final Commissioning Plan.
- A complete record of all prefunctional tests, equipment startups, and functional performance tests. Typically, the checklists used to guide and document these activities, properly

---

5. ASHRAE (1996, Section 8.3)
6. ASHRAE (1996, Section 8.7.3)

dated and signed, provide this record. It would include records of initial failures, along with the results of follow-up tests after corrective action had been taken.

- Recommendations for acceptance, or otherwise, or the systems, or conditions, that must be met before acceptance can be recommended.

Complete the Systems Manual and Record Documentation

The second major acceptance phase task is completing the HVAC systems manual and providing this document in hard copy or on CD. The components of the HVAC systems manual include:

1. The as-built or record plans and specifications, with all submittals except those that are placed in the operations and maintenance (O&M) manual.

2. The operations and maintenance manual, which is a key document. For each piece of equipment, the manual needs to include a contact name (in case of troubles), submittal and product data (all approved submittal data, cut sheets, and appropriate shop drawings), and operation and maintenance instructions.

3. The Commissioning Report,[7] which should prescribe both archival documentation (such as owner's intent) and the information the operators use. The latter category is the O&M manual outlined above. The former category is a matter of agreement between the A/E and the owner on what the owner wants to have on file.

4. A training videotape of the operating sequence for each piece of equipment is a very useful addition to the system manual. It gives easy-to-understand guidance as to system operation and procedures.

5. Manufacturers' and contractors' warranties. It is important that the owner and commissioning authority establish the warranty start and end date and document any problems that appear during that period for correction prior to warranty expiry.

Train the Operation and Maintenance Staff

The operation and maintenance staff should be adequately trained to operate the GSHP system as installed in the building.[8] The training should take place prior to occupancy of the building and be conducted by people knowledgeable with the equipment, including the designer, the contractors, and manufacturer's representative. A system operation manual should be prepared for this purpose. A videotape of the operating sequence for each piece of equipment in the system

---

7. ASHRAE (1996, Section 12); PECI (1998, 17100, Section 3.8.B).
8. ASHRAE (1996, Section 11); PECI (1998, Section 3.9).

Sidebar 2I: Operator and Maintenance Staff Training

1. Discussion of the design intent.
2. Descriptions of all equipment and systems installed and how they work.
3. Health and safety aspects.
4. A review of the Systems Manual including
   - as-built plans and specifications,
   - operation and maintenance manuals,
   - manufacturers' test data,
   - test and balance reports,
   - the functional performance test report and other material.
5. Review of all aspects of startup, operation in normal and emergency modes, shutdown procedures, seasonal changeover, and manual/automatic control.
6. How to program system controllers and thermostats.
7. Hands-on operation of equipment and systems.
8. Special tools needed and spare parts inventory.
9. Operation and adjustment of dampers, valves and controls.
10. Troubleshooting information.
11. Warranties.
12. Maintenance requirements and schedules.
13. A walk-through of the building.

---

from startup through normal operation and shutdown has been found to be very helpful as a permanent record to have on site.

Topics to cover in training are shown in Sidebar 2I.

Acceptance Phase Summary

Observation during the acceptance phase is that all test results are complete and that the system is operating as intended. It includes ensuring that the commissioning report is completed and that all system documentation—HVAC system manual, operations and maintenance manual, as-built records, training videotape—are turned over to the owner.

SUMMARY

The key to simple but adequate commissioning is adopting a simple HVAC design architecture. GSHP systems use relatively large numbers of small-capacity heat pumps and typically have many units of the same models involved in the project. Control systems can also be relatively simple, while ensuring adequate ventilation for indoor air quality (IAQ).

The *design phase* of GSHP commissioning requires a thorough site survey and characterization, accurate load modeling, and ensuring that the design chosen (and its documentation) will meet design intent.

The *construction phase* is dominated by observation of installation and verification of prefunctional checks and tests. It also involves planning, training development, and other activities to help the future building operators understand the HVAC system.

Acceptance phase activities start with functional tests and verification of all test results. It continues with full documentation: completing the Commissioning Report to include records of design changes and all as-built plans and documents and completing the O&M manual and system manual. Finally, training of the owner's operating staff occurs during the acceptance phase, after TAB is complete. Acceptance phase ends when "substantial completion" is reached. The warranty period begins from this date. It is a good idea to have a pre-warranty expiration meeting and inspection to document any problems for correction.

Chapter 3
Preventive Maintenance

INTRODUCTION

The purpose of preventive maintenance is to avoid premature equipment failures and the associated inconvenience, disruptions in building use, possible shortening of equipment service life, and extra costs associated with all of the foregoing. Maintenance and repairs based on equipment failure alone are not desirable for the following reasons:

- Equipment failure in one component can cause the subsequent failure of other items.
- Failures cause greater inconvenience for occupants and tenants than scheduled maintenance.
- Labor rates in emergency situations may be higher.
- Failed equipment may have safety or health implications.
- Failure-based maintenance is difficult to budget.

In addition to preventing failure, preventive maintenance ensures the optimum operational efficiency of equipment, leading to lower energy consumption, better indoor air quality, and lower noise levels.

SCOPE OF PREVENTIVE MAINTENANCE

It is important that the extent of a preventive maintenance program be properly determined and documented. The scope is critical since an overly extensive program will waste resources, whereas too small a scheme may not be effective in preventing failure.[1]

Preventive maintenance schedules are given in this chapter. They are intended to be used in conjunction with manufacturer's requirements and recommendations in order to achieve a balanced preventive maintenance program. Note that some maintenance steps involve the use of hazardous materials. All safety and regulatory procedures should be followed. The following maintenance schedules are presented in this chapter:

---

1. We believe that this is a major factor in the observed tenfold and greater variability in maintenance costs among buildings.

- Heat pump unit—water-to-air
- Heat pump unit—water-to-water
- Fan-coil unit
- Ground loop system
- Groundwater system

The maintenance procedure and schedule for each piece of equipment should be documented. Itemizing each maintenance task in a checklist format is a good approach. The maintenance program should also be set up to facilitate the recording of all maintenance work carried out, plus documentation of problems encountered, repairs carried out, etc.

Note: This section suggests preventive maintenance schedules for classes of equipment and some system components. Consider these as "checklists" showing items that should be evaluated for inclusion and frequency. Use manufacturers' requirements and recommendations in preference to these "defaults" wherever possible. Resist the temptation to specify more maintenance than the manufacturer requires, as it is likely to increase cost without commensurate benefits. Low maintenance is one of the advantages of GSHP designs and should not be defeated by unnecessary maintenance activities.

Table 3-1
Heat Pump Unit (Water-to-Air) Maintenance

Frequency	Maintenance Action
Quarterly	1. Check hose connections for leaks. Repair if leaking.
	2. Check unit for excessive vibration and noise.
	3. Check fan belt for wear, tension, and misalignment. Repair/adjust as required.
	4. Lubricate fan and motor bearings, if so provided and only as required.
	5. Change air filter.
	6. Cycle reversing valve to ensure operation. Repair/replace if required.
Semi-Annually	Quarterly maintenance actions, plus:
	7. Check fan and plenum for dirt/mold accumulation. Clean if required.
	8. Clean air-side coil, internal insulation, walls of dirt/mold.
	9. Clean condensate drip pan and its drain piping. Use algaecide to minimize slime.
Annually	Quarterly and semi-annual maintenance actions, plus:
	10. Check operation of contactors, relays, and reset switches. Adjust/repair/replace as needed.
	11. Check tightness of wiring connections in control panel. Tighten as needed.
	12. Check fan for bent blades and imbalance. Correct as required by reference to operating chart for pressure and temperature.
	13. Check proper operation of two-way valves, if so equipped, while checking reversing valves.

Table 3-2
Heat Pump Unit (Water-to-Water) Maintenance

Frequency	Maintenance Action
Quarterly	1. Check hose connections for leaks. Repair if leaking.
	2. Check unit for excessive vibration and noise.
	3. Cycle reversing valve to ensure operation. Repair/replace if required.
Annually	Quarterly maintenance actions, plus:
	4. Check operation of contractors, relays, and reset switches. Adjust/repair/replace as needed.
	5. Check tightness of wiring connections in control panel. Tighten as needed.

Table 3-3
Fan Coil Unit Maintenance

Frequency	Maintenance Action
Quarterly	1. Check hose connections for leaks. Repair if leaking.
	2. Check unit for excessive vibration and noise.
	3. Check fan belt for wear, tension, and misalignment. Adjust/replace as required.
	4. Lubricate fan and motor bearings, if applicable and only as required.
	5. Change air filter.
	6. Check coil and piping for leaks, corrosion, or damage. Repair/clean as needed.
Semi-Annually	Quarterly maintenance actions, plus:
	8. Check fan and plenum for dirt/mold accumulation. Clean if required.
	9. Clean coil, internal insulation, walls of dirt/mold.
	10. Clean condensate drip pan and its drain piping.
Annually	Quarterly and semi-annual maintenance actions, plus:
	11. Check tightness of wiring connections in control panel. Tighten as needed.
	12. Check fan for bent blades and imbalance. Correct as required.

Table 3-4
Ground Loop System Maintenance

Frequency	Maintenance Action
Quarterly	1. Ground loop: check static pressure reading using pete's plugs at circuit termination at mechanical room manifold.
	2. Ground loop: check antifreeze, inhibitors concentration, pH and conductivity, if applicable.
	3. Circulation pump: Check pressure difference between suction and discharge.
	4. Check makeup water use as measured by the makeup water meter to account for possible piping or earth loop leaks.
Semi-Annually	Quarterly maintenance actions, plus:
	5. Circulation pump: check suction and discharge piping and mechanical seals for leaks. Replace if required.
	6. Circulation pump: check for excessive vibration. Check motor for noise and overheating. Correct as required.
	7. Circulation pump: check alignment of pump and motor. Realign as needed.
	8. Circulation pump: lubricate motor and pump.
Annually	Quarterly and semi-annual maintenance actions, plus:
	9. Circulation pump: clean strainer.
	10. Circulation pump: test operation of standby unit load-sharing controls.

Table 3-5
Groundwater System Maintenance

Frequency	Maintenance Action
Daily	1. Check pressure drop or groundwater strainer.
	2. Check plate heat exchanger entering and leaving temperatures.
Quarterly	3. Check and record submersible pump's amperage and voltage relative to the manufacturer's specifications.
	4. Clean sediment screen.
	5. Check and record dynamic water level of production well.
	6. Check and record static water level of production well.
	7. Check and record injection pressure (pressure of water returning to the injection well).
	8. Check and record dynamic water level of injection well for impending overflow.
Semi-Annually (Spring and Fall)	Daily and quarterly maintenance actions, plus:
	9. Clean plate-frame heat exchanger with weak acid solution[a] only as temperatures indicate fouling.
	10. Reverse flow of system (if ATES system).
Annually	Daily, quarterly, and semi-annual maintenance actions, plus:
	11. Check and record submersible pump yield under normal operating conditions.
	12. Calculate pumping efficiency (efficiency should not have declined more than 15% from initial test) and record.
Whenever submersible pump or injection pipe is removed	13. Video inspection of production or injection well, particularly if water level indicates plugging is taking place.

a. This is a situation in which maintenance checklists are very useful. Plate-frame should be checked more frequently for new installations. Experience gained with the groundwater at the specific site can be used to set tear-down and cleaning interval, since groundwater chemistry is unlikely to change significantly.

Chapter 4
Troubleshooting Guide

INTRODUCTION

The scope of the troubleshooting guide covers the system as a whole, including the building distribution system, the ground installation, and a typical GSHP heat pump unit. The troubleshooting guide is subdivided into four major systems. These are:

- Heat pump unit (Table 4-1)
- Distribution system (Table 4-2)
- Ground heat exchanger (Table 4-3)
- Groundwater system (Table 4-4)

Tables 4-1 through 4-4 present a comprehensive troubleshooting guide.

HOW TO USE THE GUIDE

In order to be more useful as a reference tool, the guide is based on the system where the problem is apparent, even if the cause is in another system. Thus, the "Possible Cause" section in the tables has up to three levels, giving an indication of dependencies. Since many problems manifest themselves as performance problems with the heat pump unit, Table 4-4 is the most extensive table. Each remedy or corrective action within each table is identified with a letter, to allow referencing within the table and from Appendix B (see section 4-4).

COMMON PROBLEMS

The troubleshooting guide is based on common problems found in the operation of GSHP systems. By system these are:

Distribution system: excessive noise from ductwork; poor air distribution; leakage from piping and pumps; insulation damage; fan-coil blower belt failure, wrong blower speed or damper setting; poor thermostat location.

Ground heat exchanger: piping and pump noise; leaks; low or no flow in loop; high pump energy consumption, low antifreeze concentration; air entrainment.

Groundwater system: low flow rate in system; low well yield; overflowing injection well; short submersible pump life; fouling of plate-frame heat exchanger.

Heat pump unit: inadequate or no cooling; inadequate or no heating; excessive noise; poor temperature regulation; control circuit failure; reduction in energy efficiency.

DETAILS OF SPECIFIC REMEDIES

Appendix B gives details and procedures for selected troubleshooting remedies that are unique to GSHP systems. These are referenced to the troubleshooting guide through the table number and remedy letter. The information in Appendix B is intended as a guide only and must not be used as a substitute for manufacturer's recommended procedures or established safe practices in each specific trade. When testing and repairing GSHP systems, all applicable safety and environmental regulations must be followed.

Table 4-1
Troubleshooting GSHP Systems—Heat Pump Unit

Symptom	Possible Cause		Remedy	#
Unit supplies inadequate cooling	Airflow rate from unit too low	Air filter clogged	Clean/replace filter	a
		Damper incorrectly set	Adjust damper to higher flow rate	b
		Supply ductwork too small	Replace with larger supply duct or speed up fan	c
		Blower RPM too low	Change pulleys or speed tap	d
		Supply ductwork leaking	Repair leaks in ductwork	e
	Water flow rate through unit too low (general)	Rust or debris in unit's strainer, if so equipped. Inhibitor level low	Clean strainer, increase inhibitor level, determine why debris accumulating in system: leaks, poor flushing, etc.	f
		Inhibitor ineffective	Change inhibitor or water treatment	g
	Water flow rate through unit too low (in groundwater systems)	Flow balancing valve incorrectly set	Reset valve, rebalance system if necessary	h
		Pump not supplying adequate pressure	Upgrade pump. Check controls if VFD in use	i
	Ground loop temperature too high	Scaling in unit's water-to-refrigerant heat exchanger	Clean or replace heat exchanger. Adopt water chemistry control program using inhibitors	j

35

Table 4-1 (Continued)
Troubleshooting GSHP Systems—Heat Pump Unit

Symptom	Possible Cause	Remedy	#
Unit supplies inadequate cooling (Continued)	Ground loop temperature too high (Continued)		
	Run-out balancing valves for ground heat exchanger incorrectly set at manifold	Rebalance run-out valves	k
	Improper grouting of boreholes	Add extra boreholes or closed circuit fluid cooler to system	l
	Ground loop inadequately sized for cooling	Add extra boreholes or fluid cooler to system	m
	Lower heating loads than design	Add fluid cooler	
	Ground heat exchanger plugged or restricted	Clean and flush exchanger (Table 4-3 item x)	
	Groundwater temperature too high (standing column well)	Add bleed for extreme weather conditions (if permitted)	n
	Low refrigerant charge	Check for leaks. If leaks, repair or replace appropriate parts	o
	Restrictive metering device	Replace metering device	p
	Thermostat improperly located	Relocate thermostat	q
	Unit undersized	Recalculate loads and replace with correctly sized unit	r
	Outside air fraction to return plenum too high	Outside air fraction set unnecessarily high	Reduce outside air fraction, but maintain indoor air quality requirement	s
	Outside air damper controls defective	Replace/repair controls	t
	Outside air damper stuck open	Loosen and lubricate damper	u
	Building envelope faulty	Check for infiltration and poor insulation using infrared camera. Correct insulation, sealing windows, etc., as necessary	w

Table 4-1 (Continued)
Troubleshooting GSHP Systems—Heat Pump Unit

Symptom	Possible Cause	Remedy	#
Unit will not cool at all	Sticking reversing valve	Service or replace reversing valve	x
	Defective reversing valve relay	Replace reversing valve relay	y
	Compressor failure	Replace compressor	z
	Water flow rate through unit too low (see Table 4-1 items f to j)		aa
	Ground loop or groundwater temperature too high (see Table 4-1 items k to n)		ab
	High-pressure lock-out: Refrigerant overcharged	Reduce refrigerant charge	ac
	High-pressure lock-out: Defective high-pressure lock-out switch	Test lock-out switch and replace if defective	ad
	Airflow rate through unit too low (see Table 4-1 items a to e)		ae
	Low-pressure lock-out: Refrigerant undercharged	Increase refrigerant charge	af
	Low-pressure lock-out: Defective low-pressure lock-out switch	Test lock-out switch and replace if defective	ag
	Airflow rate from unit too low (see Table 4-1 items a to e)		ah
	Water flow rate through unit too low (see Table 4-1 items f to j)		ai
Unit supplies inadequate heating	Ground loop temperature too low: Run-out balancing valves for ground heat exchanger incorrectly set at manifold	Rebalance run-out valves	aj
	Ground loop temperature too low: Improper grouting of boreholes	Add extra boreholes or boiler to system	ak
		Add supplementary heat	al
	Ground loop inadequately sized for heating	Add extra boreholes or boiler to system	am
		Add supplementary heat	An
	Heat exchanger plugged	Flush loop (Table 4-3 item x)	

Table 4-1 (Continued)
Troubleshooting GSHP Systems—Heat Pump Unit

Symptom	Possible Cause	Remedy	#
	Groundwater temperature too low (standing column well)	Add bleed for extreme weather conditions (if permitted)	ao
	Low refrigerant charge	Check for leaks. If leaks, repair or replace appropriate parts	ap
	Restrictive metering device	Replace metering device	aq
	Thermostat improperly located	Relocate thermostat	ar
Unit supplies inadequate heating (Continued)	Unit undersized	Recalculate loads and replace with correctly sized unit	as
	Defective compressor valves	Replace compressor	at
	Outside air fraction to return plenum too high (see Table 4-1 items s to u)		au
	High ventilation load	Add energy recovery ventilation (ERV)	
	Building envelope faulty	Check for infiltration and poor insulation using infrared camera. Correct insulation, sealing windows, etc., as necessary	av
	Sticking reversing valve	Service or replace reversing valve	aw
Unit will not heat at all	Defective reversing valve relay	Replace reversing valve relay	ax
	Compressor failure	Replace compressor	ay

Table 4-1 (Continued)
Troubleshooting GSHP Systems—Heat Pump Unit

Symptom	Possible Cause	Remedy	#
Unit will not heat at all (Continued)	Low-pressure lock-out — Water flow rate through unit too low (see Table 4-1 items *f* to *j*)	Identify/correct reason for low flow and replace heat pump	az
	Heat exchanger frozen		
	Ground loop or groundwater temperature too low (see Table 4-1 items *aj* to *an*)		ba
	Refrigerant undercharged	Increase refrigerant charge	bb
	Defective low-pressure lock-out switch	Test lock-out switch and replace if defective	bc
	High-pressure lock-out — Airflow rate through unit too low (see Table 4-1 items *a* to *e*)		bd
	Air temperature entering coil too low		be
	Refrigerant overcharged	Reduce refrigerant charge	bf
	Defective high pressure lock-out switch	Test lock-out switch and replace if defective	
Excessive noise	Sound not properly attenuated	Add sound-absorbing duct liner to supply ductwork	
		Add or change vibration isolators on heat pump unit hangers	
		Add return ductwork section with sound-absorbing liner (noise isolation boot)	bg
		Add/repair vibration isolator collars on ductwork at heat pump unit connections	bh
		Relocate return air grille in suspended ceiling away from unit	bi

Table 4-1 (Continued)
Troubleshooting GSHP Systems—Heat Pump Unit

Symptom	Possible Cause		Remedy	#
Excessive noise (Continued)	Sound being amplified		Check for an item (ceiling hanger wire, etc.) that may be touching unit. Relocate or isolate from unit.	bj
	Unit in sound sensitive location		Relocate unit	bk
	Flow excessively high (water noise)		Reduce flow rate through unit	bl
Poor temperature regulation	Thermostat improperly located		Move thermostat away from: sunlight, diffuser downwash (or change diffuser flow pattern), lights, heat rejecting equipment, and sources of magnetic interference	bm
	Thermostat defective		Replace thermostat	bn
Control circuit failure	Interference from valve actuators		Check actuators to see if they are those specified, replace if not	bo
Deteriorating energy efficiency	Open-loop system (direct)	High CaCO$_3$, pH content of well water	Separate groundwater flow by adding plate-frame heat exchanger to system and institute maintenance program	bp
	Closed-loop system	Makeup water has high CaCO$_3$ content and makeup water flow rate too high	See Table 4-4 item c	bq
	Scale buildup in water-to-refrigerant heat exchanger	Inhibitor/scale-conditioner level too low	Correct inhibitor/scale-conditioner level	br

40

Table 4-2

Troubleshooting GSHP Systems—Distribution System

Symptom	Possible Cause	Remedy	#
Excessive noise from ductwork	Balancing damper incorrectly set	Adjust damper to reduce flow in branch	a
	Duct too small	Replace with duct of larger cross-sectional area	b
	Badly balanced fan	Balance fan	
	Leaks in ducts	Fix leaks in ducts	
	Lacking turning vanes	Add turning vanes	
	Fan running too fast	Slow down fan	
Inadequate supply airflow	Balancing damper incorrectly set (systems with more than one zone per heat pump)	Rebalance damper	c
	Heat pump/fan-coil unit filter clogged	Replace filter	d
	Diffuser damper incorrectly set	Adjust diffuser damper	e
	Duct too small	Replace with duct of larger cross-sectional area	f
	Fan running too slow	Speed up fan	
Poor heating/cooling distribution	Return air restricted	Undercut doors (largely applicable to residential and other light frame structures)	g
		Add door grilles	h
		Add return air grilles to ceiling plenum	i
		Add return air ducts	j
		Increase cross-sectional area of return air duct	k
Inadequate return airflow	Return air fan undersized (central distribution system)	Replace belt pulleys for higher speed or upgrade to larger fan	l

Table 4-2 (Continued)
Troubleshooting GSHP Systems—Distribution System

Symptom	Possible Cause		Remedy	#
Poor heating/cooling distribution (Continued)	Inadequate exhaust airflow	Exhaust air restricted	Add door grilles to washroom doors	m
		Exhaust air fan undersized	Replace belt pulleys for higher speed or upgrade to larger fan	n
Damage or loss of pipe insulation	Leaks from pipe		Repair pipe, replace or add insulation if insulation damage has caused condensation	o
Makeup liquid alarm on	Leak in loop valve or pipe		Repair pipe or valve	p
	Blower belt failed		Check pulley alignment, replace belt	q
		Air filter clogged	Clean/replace filter	r
	Airflow rate from unit too low	Damper incorrectly set system with multiple zones per heat pump	Adjust damper to higher flow rate	s
		Supply ductwork too small	Replace with larger supply duct	t
		Blower RPM too low	Change pulleys	u
Fan coil unit heats/cools inadequately		Supply ductwork leaking	Repair leaks in ductwork	v
	Thermostat improperly located		Relocate thermostat. Move thermostat away from: sunlight, diffuser downwash (or change diffuser flow pattern), lights, heat rejecting equipment, and sources of magnetic interference	w
	Unit undersized		Recalculate loads and replace with correctly sized unit	x

Table 4-2 (Continued)
Troubleshooting GSHP Systems—Distribution System

Symptom	Possible Cause			Remedy	#
	Pump leaks			Replace seals	y
		Fitting or connection insufficiently tight		Tighten connection/fitting	z
		Crack in piping		Repair/replace section of pipe	aa
Leaks in system	Piping/fittings leaks	Pipe corrosion	Low surface tension. System charged with potassium acetate	Replace with antifreeze solution of higher surface tension	ab
			Ethanol or methanol antifreeze	Add/increase inhibitor	ac
				Replace ethanol with propylene glycol antifreeze	ad
	"Leak" is actually condensation			Insulate pipe c/w vapor barrier	af
	Closed fire damper in ductwork			Check and fix	
	Failed fan motor			Replace fan motor	
No airflow	Control circuit failure			Check and fix	
	Condensate high level switch activated			Check and fix	
	Tripped circuit breaker			Check and rectify	
	Damaged power line			Check/repair wiring	

Table 4-3

Troubleshooting GSHP Systems—Ground Heat Exchanger

Symptom	Possible Cause		Remedy	#
Noise from piping	Air in ground loop	Improper system startup or flushing	Re-start or re-flush using correct procedure	a
		Leak in ground loop system	Find and repair leak	b
		Loose connections in valves and other pipe fittings	Tighten valves and fittings	c
		Inadequate air separation in system	Add air separator	d
			Add expansion tank	e
	Defective circulating pump		Repair or replace circulating pump	f
Loss of pressure	Leak in ground loop or associated piping		Evaluate location of leak and repair (see Appendix B, Section B.2 and B.3)	g
	Leak in distribution system (if integral with ground loop)		See Table 4-2 items y to ae	h
	Relaxation of plastic pipe		Normal	i
	Change of fluid temperature		Normal (within 10-30 psi range)	j
	Leak in header, valve or fitting	Loose mechanical connection	Tighten fitting or replace defective part	k
		Poor fusion weld in HDPE piping	Repair (see Appendix B, Section B.3)	l
		Puncture or defect in HDPE piping	Repair (see Appendix B, Section B.3)	m
Water dripping from mechanical room piping	Leak		See Table 4-2 items y to ae and Table 4-3 items k to m	n
	Condensation		Insulate pipe c/w vapor barrier	o
Ground above borefield saturated	Leak in ground loop or associated piping		Find and repair leak (see Appendix B, Section B.2 and B.3)	p

Table 4-3 (Continued)
Troubleshooting GSHP Systems—Ground Heat Exchanger

Symptom	Possible Cause		Remedy	#
No flow in loop	No power to pump	Fuse/breaker tripped	Reset fuse Repair or replace VFD	q
		Improperly wired or phase loss	Rewire correctly	r
	Pump shaft stuck or jammed		Check for obstructions and remove	s
	Air lock in system		Flush loop to remove air (see Appendix B, Section B.1)	t
	Improperly sized pump		Replace with correctly sized pump	u
	Defective pump		Replace pump	v
	Line strainer full		Clean out line strainer	w
Low flow in loop	Sediment buildup in loop		Flush loop (see Appendix B, Section B.1)	x
	Loop fluid freezing		Set thermostat to cooling to defrost loop. Upgrade antifreeze concentration.	y
	Kink in loop piping		Locate kink and replace section (see Appendix B, Section B.3)	z
	Leak in system		Find and repair leak (see Appendix B, Section B.2 and B.3)	aa
Improper flow	Three-phase motor turning in wrong direction		Check rotation and reverse phase connections	

Table 4-3 (Continued)
Troubleshooting GSHP Systems—Ground Heat Exchanger

Symptom	Possible Cause	Remedy	#
Excessive noise from pumps	Air in system	See Table 4-3 items *a* to *e*	ab
	Inadequate vibration/noise attenuation	Add pump mount vibration isolators, in-line pipe vibration dampers, etc.	ac
		Add acoustic insulation	ad
	Viscosity of loop fluid antifreeze solution too high	Reduce concentration of antifreeze to the minimum required for freeze protection	ae
		Switch to a lower viscosity antifreeze	af
High pump energy consumption	Pressure drop of actual system exceeds that predicted for the design	Replace pump and/or motor with one matched to the actual pressure and flow requirements	ag
	Building setback controls incorrectly set	Adjust to appropriate setback positions	ah
	Pumping flow rate exceeds thermal load requirement most of the time	Add variable-speed drive to pump	ai
		Add borefield bypass loop and controls	aj
Low antifreee concentration	Leak in system (no antifreeze makeup system)	Repair header, valves or loop (see Appendix B, Section B.3)	ak
Air in system	See Table 4-3 items *a* to *e*		al
Makeup water/antifreeze alarm on	Leak in header or ground loop	Repair header or loop (see Appendix B, Section B.3)	am
	Faulty automatic air vents (leaking)	Replace/repair automatic air vent or replace with manual air vents	an

Table 4-4

Troubleshooting GSHP Systems—Groundwater System

Symptom	Possible Cause		Remedy	#
Low flow rate through well system	Heat exchanger or well screen encrusted or clogged		Redevelop production well (see Appendix B, Section B.4)	a
		Iron bacteria fouling	Check Fe^{2+} content of well water. Iron bacteria growth is likely if: $Fe^{2+} > 0.2$ ppm; water temp. 46°F to 61°F; dissolved $O_2 < 5$ ppm; and pH = 6 to 8. Clean (see Appendix B, Section B.6) and apply appropriate water conditioning.	b
		Scaling	Check if maps (such as Rafferty 1999) indicate > 200 ppm $CaCO_3$ hardness in region. If so, scaling likely. Test if unsure. Clean heat exchanger (see Appendix B, Section B.6). Consider a regularly scheduled weak acid cleaning of the heat exchanger to prevent significant scale buildup.	c
	Groundwater strainer full		Clean or blowdown	
	Submersible pump performance poor	Low voltage due to poor wiring leading to pump	Upgrade wires to pump	d

47

Table 4-4 (Continued)
Troubleshooting GSHP Systems—Groundwater System

Symptom	Possible Cause	Remedy	#
Overflowing injection well	Injection well plugged	Redevelop injection well (see Appendix B, Section B.4)	e
		Install surface separator (strainer) on production well	f
		Reevaluate pumping needs to see if well flow rate can be reduced or give preference to lower sand well in multiple well system,	g
	Sand/particles in discharge flow (> 5 ppm)	Improve grade of production well inlet screen	h
		Improve production well gravel pack, where justifiable	i
		Increase area of injection well screen, where justifiable	j
		Convert to system with seasonal reversal of flow direction	k
	Iron bacteria fouling of injection well screen	See Table 4-4 item b	m
	Inadequate injection capacity in design	Add extra injection well	n
Pump short cycles (<15 min cycles)	Setpoint range too small	Increase setpoint range (see Appendix B, Section B.7)	o
	Too much pump capacity	Decrease pump capacity if possible	p
		Add closed storage tank to system	q

Table 4-4 (Continued)
Troubleshooting GSHP Systems—Groundwater System

Symptom	Possible Cause		Remedy	#
Submersible pump motor failure	Pump short cycled (<15 min cycles)		See Table 4-4 items *o* to *q*. Remove and replace pump/motor (Appendix B, Section B.5).	r
	Motor damaged by lightning because of inadequate ground for building electrical system		Resistance to ground should be less than 25 ohms	
	Inadequate pump cooling	Pump cooling shroud dislodged or damaged	Remove and replace pump/motor (see Appendix B, Section B.5). Ensure that new motor has a properly designed cooling shroud.	s
	Too much sand in system		See Table 4-4 items *g* to *i*. Remove and replace pump/motor (see Appendix B, Section B.5).	t
Corrosion of plate-frame heat exchanger	Well water chloride level > 140 ppm, but less than 400 ppm		Replace with 316 stainless steel heat exchanger	u
	Well water chloride level > 400 ppm		Replace with titanium heat exchanger	v

49

Chapter 5
References

ASHRAE. 1996. *ASHRAE Guideline 1-1996, The HVAC Commissioning Process*. Atlanta: American Society of Heating, Refrigerating and Air-Conditioning Engineers, Inc.

ASHRAE. 1999. *1999 ASHRAE Handbook—HVAC Applications*. Atlanta: American Society of Heating, Refrigerating and Air-Conditioning Engineers, Inc.

ASHRAE. 1995. *Commercial/Institutional Ground-Source Heat Pump Engineering Manual*. Altanta: American Society of Heating, Refrigerating and Air-Conditioning Engineers.

CMHC. 1996. *The Design of Mechanical and Electrical Systems in Multi-Unit Residential Buildings*. Canada Mortgage and Housing Corporation.

Cane, D., and J. Garnet. 1999. Update on maintenance and service costs of commercial building ground-source heat pump systems. *ASHRAE Transactions* 106(1).

Cane, D., B. Clemes, and A. Morrison. 1996. Operating experiences with commercial ground-source heat pumps—Part 1. *ASHRAE Tranactions* 102(1).

Cane, D., A. Morrison, and C. Ireland. 1998. Operating experiences with commercial ground-source heat pumps—Part 2. *ASHRAE Transactions* 104(2).

Coad, W. 1996. Indoor air quality—A design parameter. *ASHRAE Journal* 38(6).

CSA C448. 2002. *Design and Installation of Earth Energy Systems*. Canadian Standards Association.

Downey, T., and J. Proctor. 1998. Investigations of Multiple Callback Situations with Residential GeoExchange Systems. Final Report, November 13, 1997. Geothermal Heat Pump Consortium Report RP-025.

Driscoll, F.G. 1986. *Groundwater and Wells*, 2d ed. St. Paul, Minnesota: Johnson Filtration Systems Inc.

Driscopipe®. Polyethylene Piping System Manual.

Elovitz, K.M. 1992. Commissioning building mechanical systems. *ASHRAE Transactions* 98(2).

EPRI. 1997. *Grouting for Vertical Geothermal Heat Pump Systems: Engineering Design and Field Procedures Manual*. Final Report, EPRI TR-109169.

GeoExchange. *Outside the Loop: A Newsletter for Geothermal Heat Pump Designers and Installers*. University of Alabama. Published quarterly.

Hornreich, M.A. 1994. Legal aspects of commissioning of building systems. *ASHRAE Transactions* 100(2).

IGSHPA. 1991. Grouting procedures for ground-source heat pump systems. Stillwater, Okla.: International Ground-Source Heat Pump Association.

IGSHPA. 1997. Closed-loop/geothermal heat pump systems—Design and installation standards. Stillwater, Okla.: International Ground-Source Heat Pump Association.

IGSHPA. 2000. Grouting for vertical geothermal heat pump systems. *Engineering Design and Field Procedures Manual*. Stillwater, Okla.: International Ground-Source Heat Pump Association.

Kavanaugh, S.P., and K. Rafferty. 1997. *Ground-Source Heat Pumps: Design of Geothermal Systems for Commercial and Institutional Buildings*. Atlanta: American Society of Heating, Refrigerating and Air-Conditioning Engineers, Inc.

Lehr, J., S. Hurlburt, B. Gallagher, and J. Voytek. 1988. *Design and Construction of Water Wells: A Guide for Engineers*. New York: Van Nostrand Reinhold. Prepared for the National Water Well Association.

McCray. 1997. *Guidelines for the Construction of Vertical Boreholes for Closed Loop Heat Pump Systems*. National Groundwater Association.

PECI. 1998. *Commissioning Guide Specifications, version 2.05*. Portland Energy Conservation, Inc.

Rafferty, K. 1999. *Scaling in Geothermal Heat Pump Systems*. Prepared for U.S. Department of Energy. Contract Number DE-FG07-90ID 13040. July.

Rafferty, K. 2000. Design aspects of commercial open-loop heat pump systems. Proceedings of the Fourth International Conference: Heat Pumps in Cold Climates, Aylmer, Quebec, Canada, Aug 17-18.

Roscoe Moss Co. 1990. *Handbook of Ground Water Development*. Wiley-Inter-Science.

Singh, J.B., P.E. Foster, Jr., and A.W. Hunt. 2000. Representative operating problems of commercial ground-source and groundwater-source heat pumps. *ASHRAE Transactions* 106(2).

Smith, S.A. 1995. *Monitoring and Remediation Wells: Problem Prevention, Maintenance, and Rehabilitation*. Boca Raton, Florida: CRC Press.

Sclairpipe®. High Density Polyethylene Pipe, Systems Design Manual

Thomas, W., and M. Madgett. 2000. *Generic Guide Specifications for Geothermal Heat Pump System Installation*. Oak Ridge National Laboratory, ORNL/TM-2000/132.

Tracer Research Corporation website: www.tracerResearch.com.

Trane. Installation Owner Diagnostics: Axiom GEHA Water-Source Comfort System Horizontal Configuration. Bulletin GEHA-SVX01B-EN.

Tseng, P.C., S.E. Batterden, and W.A. Appenzellar. 1994. Assessment criteria for commissioning and construction quality control—Assessing the effectiveness of total building commissioning. *ASHRAE Transactions* 100(2).

Tyler, R.J. 1994. Legal aspects of building commissioning—Claim avoidace. *ASHRAE Transactions* 100(2).

Appendix A
Specification Guide for Commissioning of GSHP Systems

APPENDIX A-1—COMMISSIONING GUIDE SPECIFICATIONS

The guide presented in this appendix assumes that the A/E commissions the system. This is based on the premise that commissioning lies within the traditional responsibility of the architect/design engineer when a disciplined design and construction program, focused on simplicity, is applied to GSHP systems. This is particularly true in small buildings. A separate commissioning authority should not be needed for these GSHP systems. However, the A/E needs to be qualified, willing, and compensated. If not, a separate commissioning authority is needed.

Commissioning Guide Specifications
Section 17100 Commissioning Requirements

The following guide specifications are intended to be reviewed and modified to meet the specific commissioning needs and requirements for the current GSHP project and systems. Any modifications to this specification shall only be made after consultation with the owner's representative and with approval of the engineer of record. Where there are check boxes or fill-in blanks, fill in as appropriate and delete all but the choice(s) that apply. Delete all direction boxes. The text presented here is simplified from the DOE/PECI version.[a] This model has the A/E or mechanical designer as the lead authority for commissioning, rather than a separate "Commissioning Authority." This is appropriate where the building is relatively small or the A/E is qualified, willing, and compensated. The content of this guide may be appropriate for large federal projects. For commercial projects and relatively small buildings, the project team may choose to simplify Section 17100 greatly.

---

a. PECI (Portland Energy Conservation, Inc. 1998. Commissioning Guide: Specifications, v. 2.05. Available from www.peci.org.

TABLE OF CONTENTS (WITH SELECTED SUBSECTIONS)

PART 1—GENERAL

1.1 Description

A. **Commissioning.** Commissioning is a systematic process of ensuring that all building systems perform according to the design intent and the owner's operational needs. This is achieved by beginning in the design phase and documenting design intent and continuing through construction, acceptance, and the warranty period with actual verification of performance. The commissioning process includes system documentation, equipment startup, control system calibration, testing and balancing, performance testing, and training.

Commissioning during the construction phase is intended to achieve the following specific objectives according to the Contract Documents:
1. Verify that applicable equipment and systems are installed according to the manufacturer's recommendations and to industry-accepted minimum standards and that they receive adequate operational checkout by installing contractors.
2. Verify and document proper performance of equipment and systems.
3. Verify that O&M documentation left on site is complete.
4. Verify that the owner's operating personnel are adequately trained.

B. The commissioning process does not take away from or reduce the responsibility of the system designers or installing contractors to provide a finished and fully functioning product.

C. **Abbreviations.** The following are common abbreviations used in the *Specifications* and in the *Commissioning Plan*. Definitions are found in Section 1.6.

A/E	=	Architect and design engineers	PC	= Prefunctional checklist
			Cx	= Commissioning
FT	=	Functional performance test	PM	= Project manager (of the owner)
GC	=	General contractor (prime)	Cx Plan	= Commissioning plan document
CC	=	Controls contractor		
MC	=	Mechanical contractor	Subs	= Subcontractors to general
CM	=	Construction manager (the owner's representative)	EC	= Electrical contractor
			TAB	= Test and balance contractor

1.2 Coordination

A. **Commissioning Team.** The commissioning team consists of the project manager (PM), the designated representative of the owner's construction management firm (CM), the general contractor (GC or Contractor), the architect and design engineers (A/E, particularly the mechanical engineer), the mechanical contractor (MC), the electrical contractor (EC), the TAB representative, the controls contractor (CC), and any other installing subcontractors or suppliers of equipment. If known, the owner's building or plant operator/engineer is also a member of the commissioning team.

B. **Management.** The A/E shall direct commissioning activities on behalf of the owner.

C. **Scheduling.** The GC will integrate all commissioning activities into the master schedule. The Commissioning Plan will provide a format for detailed schedules.

1.3 Commissioning Process

> If no commissioning plan was issued at bidding, delete the first sentence of the following paragraph.

A. **Commissioning Plan.** The Commissioning Plan, Draft 2, provided as part of the bid documents, is binding on the Contractor. The Commissioning Plan provides guidance in the execution of the commissioning process. The *Specifications* will take precedence over the *Commissioning Plan*.

B. **Commissioning Process.** The following narrative provides a brief overview of the typical commissioning tasks during construction and the general order in which they occur.

    1. Commissioning during construction begins with a scoping meeting where the commissioning process is reviewed with the commissioning team members.

    2. Additional meetings will be required throughout construction—to plan, scope, coordinate, schedule future activities, and resolve problems.

    3. Equipment documentation is submitted during normal submittals, including detailed startup procedures.

    4. In general, the checkout and performance verification proceeds from simple to complex, from component level to equipment to systems and intersystem levels with prefunctional checklists being completed before functional testing.

    5. The Subs, under their own direction, execute and document the prefunctional checklists and perform startup and initial checkout. The A/E documents that the checklists and startup were completed according to the approved plans and may witness startup of selected equipment.

6. The A/E develops specific equipment and system functional performance test procedures. The Subs review the procedures.
7. The procedures are executed by the Subs, under the direction of, and documented by, the A/E.
8. Items of noncompliance in material, installation, or setup are corrected at the Sub's expense and the system retested.
9. The A/E reviews the O&M documentation for completeness.
10. Commissioning is completed before Substantial Completion.
11. The A/E reviews, pre-approves, and coordinates the training provided by the Subs and verifies that it was completed.
12. Deferred testing is conducted, as specified or required.

1.4 Related Work

A. Specific commissioning requirements are given in the following sections of these specifications. All of the following sections apply to the work of this section.

00800 Supplementary Conditions	Provides for a penalty if commissioning is not completed by the Functional Completion milestone.
01040a Coordination	Introduces commissioning and refers to Division 17.
01300 Submittals	Alerts all parties that additional detail in submittals may be required and directs to Division 17.
01700 Project Close-out	Defines substantial completion and functional completion milestones, relative to commissioning.
01730 O&M Data	Alerts all parties that O&M documentation may be more detailed and directs to Division 17.
15010 Mechanical General	Alerts the mechanical contractor to Cx responsibilities in 15995.
15950 Automatic Controls	Lists special requirements and alerts the controls contractor of the special requirements of the control contractor and control system in 15995.
15990 TAB	Alerts the TAB of Cx responsibilities in 15995.

15995 Mechanical Cx	Describes the Cx responsibilities of the mechanical, controls, and TAB contractors and the prefunctional testing and startup responsibilities of each. Points to 15997 for functional testing requirements.
15997 Mechanical Testing Requirements	Describes the specific functional testing requirements for Division 15 equipment in the project.
15998 Mechanical Prefunctional Checklists	Provides the prefunctional checklists for use on this project, including items for Division 15 *and* Division 16.
15999 Mechanical Functional Tests—Examples	Provides example functional test procedures and formats for mechanical equipment.
16010 Electrical General	Alerts the electrical contractor of Cx responsibilities in 16995.
16995 Electrical Cx	Describes the Cx responsibilities of the electrical contractor.
16997 Electrical Testing	Describes the specific functional testing requirements.
16998 Electrical Prefunctional Checklists	Points to Section 15998.
16999 Electrical Functional Tests—Examples	Provides example functional test procedures and formats for electrical equipment.
17100 Commissioning	Describes the commissioning process, responsibilities common to all parties, responsibilities of the A/E, CM, PM, GC, and suppliers. The unique MC, CC, TAB, and EC responsibilities are included in Division 15 and 16.

List other sections or divisions that have systems commissioned. This may include division 10000 (general), division 20000 (site), and division 30000 (grout and backfill) for aspects of ground-coupled, surface water, or groundwater heat exchanger installation. Include alerts in Section 10 of the division, division commissioning responsibilities in section 995 of the division, and actual test requirements in section 997, all similar to Division 15.

1.5 Responsibilities

A. The responsibilities of various parties in the commissioning process are provided in this section. The responsibilities of the mechanical contractor, TAB, and controls contractor are in Division 15 and those of the electrical contractor in Division 16 and those of [list other sections where requirements of other divisions are found]. It is noted that the services for the project manager, construction manager, architect, HVAC mechanical and electrical designers/engineers, and commissioning authority are not provided for in this contract. That is, the contractor is not responsible for providing their services. Their responsibilities are listed here to clarify the commissioning process.

Specification writer should make sure that language in the contract and general conditions states that the contractors are paying for their part in the commissioning and testing; the owner is not paying as an extra.

If no commissioning plan was issued at bidding, delete item number 1 under B, "Follow the commissioning plan."

B. All Parties

1. Follow the commissioning plan.
2. Attend commissioning scoping meeting and additional meetings, as necessary.

C. Architect (of A/E)

Construction and Acceptance Phase

1. Attend the commissioning scoping meeting and selected commissioning team meetings.
2. Perform normal submittal review, construction observation, as-built drawing preparation, O&M manual preparation, etc., as contracted.
3. Provide any design narrative documentation needed.
4. Coordinate resolution of system deficiencies identified during commissioning, according to the Contract Documents.
5. Prepare and submit final as-built design intent documentation for inclusion in the O&M manuals. Review and approve the O&M manuals.

Warranty Period

1. Coordinate resolution of design nonconformance and design deficiencies identified during warranty-period commissioning.

D. Mechanical and Electrical Designers/Engineers (of the A/E)

Construction and Acceptance Phase

1. Perform normal submittal review, construction observation, as-built drawing preparation, etc., as contracted. On-site observation should be completed just prior to system startup.

2. Provide any design narrative and sequences documentation needed. The designers shall assist (along with the contractors) in clarifying the operation and control of commissioned equipment in areas where the specifications, control drawings, or equipment documentation is not sufficient for writing detailed testing procedures.

3. Attend commissioning scoping meetings and other selected commissioning team meetings.

4. Participate in the resolution of system deficiencies identified during commissioning, according to the Contract Documents.

5. Prepare and submit the final as-built design intent and operating parameters documentation for inclusion in the O&M manuals. Review and approve the O&M manuals.

6. From the Contractor's red-line drawings, edit and update one-line diagrams developed as part of the design narrative documentation and those provided by the vendor as shop drawings for the chilled and hot water, condenser water, domestic water, steam, and condensate systems, supply, return, and exhaust air systems, and emergency power system.

Optionally, the Contractor may be specified to update the one-line drawings.

---

7. Provide a presentation at one of the training sessions for the Owner's personnel.

OPTIONAL:

8. __ Review, __ Approve the prefunctional checklists for major pieces of equipment for sufficiency prior to their use.

9. __ Review, __ Approve the functional test procedure forms for major pieces of equipment for sufficiency prior to their use.

10. __ Witness testing of selected pieces of equipment and systems:

_____

_____.

Warranty Period

1. Participate in the resolution of noncompliance, nonconformance, and design deficiencies identified during commissioning during warranty-period commissioning.

If no commissioning plan was issued at bidding, modify as appropriate, any references to the commissioning plan throughout the rest of this section.

E. Construction Manager—Owner's Representative (CM)

Construction and Acceptance Phase

1. Facilitate the coordination of the commissioning work by the A/E, and, with the GC, ensure that commissioning activities are being scheduled into the master schedule.
2. Review and approve the final *Commissioning Plan—Construction Phase*.
3. Attend a commissioning scoping meeting and other commissioning team meetings.
4. Perform the normal review of Contractor submittals.
5. Furnish a copy of all construction documents, addenda, change orders, and approved submittals and shop drawings related to commissioned equipment.
6. Review and approve the functional performance test procedures submitted by the A/E prior to testing.
7. When necessary, observe and witness prefunctional checklists, startup, and functional testing of selected equipment.
8. Review commissioning progress and deficiency reports.
9. Coordinate the resolution of noncompliance and design deficiencies identified in all phases of commissioning.
10. Sign-off (final approval) on individual commissioning tests as completed and passing. Recommend completion of the commissioning process to the Project Manager.
11. Assist the GC in coordinating the training of owner personnel.

Warranty Period

1. Assist the A/E as necessary in the seasonal or deferred testing and deficiency corrections required by the specifications.

F. Owner's Project Manager (PM)

Construction and Acceptance Phase

1. Manage the contract of the A/E and of the GC.
2. Arrange for facility operating and maintenance personnel to attend various field commissioning activities and field training sessions according to the *Commissioning Plan—Construction Phase.*
3. Provide final approval for the completion of the commissioning work.

Warranty Period

1. Ensure that any seasonal or deferred testing and any deficiency issues are addressed.

G. General Contractor (GC)

Construction and Acceptance Phase
1. Facilitate the coordination of the commissioning work, and ensure that commissioning activities are being scheduled into the master schedule.
2. Include the cost of commissioning in the total contract price.
3. Furnish a copy of all construction documents, addenda, change orders, and approved submittals and shop drawings related to commissioned equipment to the A/E.
4. In each purchase order or subcontract written, include requirements for submittal data, O&M data, commissioning tasks, and training.
5. Ensure that all Subs execute their commissioning responsibilities according to the Contract Documents and schedule.
6. A representative shall attend a commissioning scoping meeting and other necessary meetings scheduled to facilitate the Cx process.
7. Coordinate the training of owner personnel.
8. Prepare O&M manuals, according to the Contract Documents, including clarifying and updating the original sequences of operation to as-built conditions.

Warranty Period
1. Ensure that Subs execute seasonal or deferred functional performance testing, witnessed by the A/E, according to the specifications.
2. Ensure that Subs correct deficiencies and make necessary adjustments to O&M manuals and as-built drawings for applicable issues identified in any seasonal testing.

H. Equipment Suppliers

1. Provide all requested submittal data, including detailed startup procedures and specific responsibilities of the Owner to keep warranties in force.
2. Assist in equipment testing according to agreements with Subs.
3. Include all special tools and instruments (only available from vendor, specific to a piece of equipment) required for testing equipment according to these Contract Documents in the base bid price to the Contractor, except for stand-alone datalogging equipment that may be used for commissioning.
4. Through the contractors they supply products to, analyze specified products and verify that the designer has specified the newest, most updated equipment reasonable for this project's scope and budget.
5. Provide information requested by A/E regarding equipment sequence of operation and testing procedures.
6. Review test procedures for equipment installed by factory representatives.

1.6 Definitions

Acceptance Phase: Phase of construction after startup and initial checkout when functional performance tests, O&M documentation review, and training occur.

Approval: Acceptance that a piece of equipment or system has been properly installed and is functioning in the tested modes according to the Contract Documents.

Architect/Engineer (A/E): The prime consultant (architect) and subconsultants who comprise the design team, generally the HVAC mechanical designer/engineer and the electrical designer/engineer.

Basis of Design: The Basis of Design is the documentation of the primary thought processes and assumptions behind design decisions that were made to meet the design intent. The basis of design describes the systems, components, conditions, and methods chosen to meet the intent. Some reiterating of the design intent may be included.

Commissioning Plan: An overall plan, developed before or after bidding, that provides the structure, schedule, and coordination planning for the commissioning process.

Contract Documents: The documents binding on parties involved in the construction of this project (drawings, specifications, change orders, amendments, contracts, Cx Plan, etc.).

Contractor: The general contractor or authorized representative.

Control System: The central building energy management control system.

Construction Manager (CM): The owner's representative in the day-to-day activities of managing construction. The general contractor reports to the CM.

Datalogging: Monitoring flows, currents, status, pressures, etc., of equipment using stand-alone dataloggers separate from the control system.

Deferred Functional Tests: Functional performance tests that are performed later, after substantial completion, due to partial occupancy, equipment, seasonal requirements, design, or other site conditions that prevent the test from being performed.

Deficiency: A condition in the installation or function of a component, piece of equipment, or system that is not in compliance with the Contract Documents (that is, does not perform properly or is not complying with the design intent).

Design Intent: A dynamic document that provides the explanation of the ideas, concepts, and criteria that are considered to be very important to the owner. It is initially the outcome of the programming and conceptual design phases.

Design Narrative or Design Documentation: Sections of either the Design Intent or Basis of Design.

Factory Testing: Testing of equipment on-site or at the factory by factory personnel with an Owner's representative present.

Functional Performance Test (FT): Test of the dynamic function and operation of equipment and systems using manual (direct observation) or monitoring methods. Functional testing is the dynamic testing of systems (rather than just components) under full operation (e.g., the chiller pump is tested interactively with the chiller functions to see if the pump ramps up and down to maintain the differential pressure setpoint). Systems are tested under various modes, such as during low cooling or heating loads, high loads, component failures, unoccupied, varying outside air temperatures, fire alarm, power failure, etc. The systems are run through all the control system's sequences of operation, and components are verified to be responding as the sequences state. Traditional air or water testing and balancing (TAB) is not functional testing, in the commissioning sense of the word. TAB's primary work is setting up the system flows and pressures as specified, while functional testing is verifying that which has already been set up. The A/E develops the functional test procedures in a sequential written form, coordinates, oversees, and documents the actual testing, which is usually performed by the installing contractor or vendor. FTs are performed after prefunctional checklists and startup are complete.

General Contractor (GC): The prime contractor for the project. Generally refers to all the GC's subcontractors as well. Also referred to as the Contractor, in some contexts.

Indirect Indicators: Indicators of a response or condition, such as a reading from a control system screen reporting a damper to be 100% closed.

Manual Test: Using hand-held instruments, immediate control system readouts or direct observation to verify performance (contrasted with analyzing monitored data taken over time to make the "observation").

Monitoring: The recording of parameters (flow, current, status, pressure, etc.) of equipment operation using dataloggers or the trending capabilities of control systems.

Noncompliance: See *Deficiency.*

Nonconformance: See *Deficiency.*

Overwritten Value: Writing over a sensor value in the control system to see the response of a system (e.g., changing the outside air temperature value from 50°F to 75°F to verify economizer operation). See also *Simulated Signal.*

Owner-Contracted Tests: Tests paid for by the Owner outside the GC's contract and outside the scope of the formal commissioning process. These tests will not be repeated during functional tests if properly documented.

Phased Commissioning: Commissioning that is completed in phases (by floors, for example) due to the size of the structure or other scheduling issues, in order to minimize the total construction time.

Prefunctional Checklist (PC): A list of items to inspect and elementary component tests to conduct to verify proper installation of equipment, provided to the Sub. Prefunctional checklists are primarily static inspections and procedures to prepare the equipment or system for initial operation (e.g., belt tension, oil levels OK, labels affixed, gages in place, sensors calibrated, etc.). However, some prefunctional checklist items entail simple testing of the function of a component, a piece of equipment, or system (such as measuring the voltage imbalance on a three-phase pump motor of a chiller system). The word *pre*functional refers to *before* functional testing. Prefunctional checklists augment and are combined with the manufacturer's startup checklist. Even without a commissioning process, contractors typically perform some, if not many, of the prefunctional checklist items the A/E will recommend. However, few contractors document in writing the execution of these checklist items. Therefore, for most equipment, the contractors execute the checklists on their own. Commissioning only requires that the procedures be documented in writing and does not witness much of the prefunctional checklisting, except for larger or more critical pieces of equipment.

Project Manager (PM): The contracting and managing authority for the Owner over the design and/or construction of the project, a staff position.

Sampling: Functionally testing only a fraction of the total number of identical or near identical pieces of equipment. Refer to Section 17100, Part 3.6, F for details.

Seasonal Performance Tests: FT that are deferred until the system(s) experience conditions closer to their design conditions.

Simulated Condition: Condition that is created for the purpose of testing the response of a system (e.g., applying a hair blower to a space sensor to see the response in a VAV box).

Simulated Signal: Disconnecting a sensor and using a signal generator to send an amperage, resistance, or pressure to the transducer and DDC system to simulate a sensor value.

Specifications: The construction specifications of the Contract Documents.

Startup: The initial starting or activating of dynamic equipment, including executing prefunctional checklists.

Subs: The subcontractors to the GC who provide and install building components and systems.

Test Procedures: The step-by-step process that must be executed to fulfill the test requirements. The test procedures are developed by the A/E.

Test Requirements: Requirements specifying what modes and functions, etc., shall be tested. The test requirements are not the detailed test procedures. The test requirements are specified in the Contract Documents (Sections 15997; 16997, etc.).

Trending: Monitoring using the building control system.

Vendor: Supplier of equipment.

Warranty Period: Warranty period for entire project, including equipment components. Warranty begins at Substantial Completion and extends for at least one year, unless specifically noted otherwise in the Contract Documents and accepted submittals.

1.7 Systems To Be Commissioned

A. The following checked systems will be commissioned in this project.

Equipment and System	Functional Test Requirements Specified In:
HVAC System	
__ Pumps	15997
__ Cooling tower	15997
__ Piping systems	15997
__ Ductwork	15997
__ Variable-frequency drives	15997
__ Air handlers	15997
__ Packaged units (AC and HP)	15997
__ Terminal units (air)	15997
__ Heat exchangers	15997
__ Computer room units	15997
__ Fume hoods	15997
__ Lab room pressures	15997
__ Specialty fans	15997
__ Testing, adjusting, and balancing work	15997
__ Chemical treatment systems	15997
__ HVAC control system	15997
__ Fire and smoke dampers	15997
__ Indoor air quality[a]	15997
__ Equipment sound control	15997
__ Equipment vibration control	15997
__ Egress pressurization	15997

a. Indoor air quality (IAQ) commissioning does not ensure that indoor air quality will be adequate or without deficiency at building turnover or during occupancy, unless the owner has specifically specified that actual air quality testing be performed. Commissioning indoor air quality entails performing tasks that minimize the potential for IAQ problems, but it does not eliminate their possibility.

Electrical System

__ Sweep or scheduled lighting controls	16997
__ Daylight dimming controls	16997
__ Lighting occupancy sensors	16997
__ Power quality	16997
__ Security system	16997
__ Emergency power system	16997
__ UPS systems	16997
__ Fire and smoke alarm	16997
__ Fire protection systems	16997
__ Communications system	16997
__ Public address/paging	16997

Other

__ Service water heaters	15997
__ Service water booster pumps	15997
__ Refrigeration systems	15997
__ Medical gas systems	15997

PART 2—PRODUCTS

2.1 Test Equipment

A. All standard testing equipment required to perform startup and initial checkout and required functional performance testing shall be provided by the division contractor for the equipment being tested. For example, the mechanical contractor of Division 15 shall ultimately be responsible for all standard testing equipment for the HVAC system and controls system in Division 15, except for equipment specific to and used by TAB in their commissioning responsibilities. Two-way radios shall be provided by the division controller.

B. Special equipment, tools, and instruments (only available from vendor, specific to a piece of equipment) required for testing equipment according to these Contract Documents shall be included in the base bid price to the contractor and left on site, except for stand-alone datalogging equipment that may be used by the A/E or subcontractors to the A/E.

C. Datalogging equipment and software required to test equipment will be provided by the A/E, but it shall not become the property of the Owner.

D. All testing equipment shall be of sufficient quality and accuracy to test and/or measure system performance with the tolerances specified in the *Specifications*. If not otherwise noted, the following minimum requirements apply: temperature sensors and digital thermometers shall have a certified calibration within the past year to an accuracy of 0.5°F and a resolution of ±0.1°F. Pressure sensors shall have an accuracy of ±2.0% of the value range being measured (not full range of meter) and have been calibrated within the last year. All equipment shall be calibrated according to the manufacturer's recommended intervals and when dropped or damaged. Calibration tags shall be affixed or certificates readily available.

E. Refer to Section 17100, Part 3.6 E for details regarding equipment that may be required to simulate required test conditions.

PART 3—EXECUTION

3.1 Meetings

A. **Scoping Meeting.** Within [60 to 90 depending on building size] days of commencement of construction, the A/E will schedule, plan, and conduct a commissioning scoping meeting with the entire commissioning team in attendance. Meeting minutes will be distributed to all parties. Information gathered from this meeting will allow the A/E to revise the Draft 2 *Commissioning Plan* to its "final" version, which will also be distributed to all parties.

B. **Miscellaneous Meetings.** Other meetings will be planned and conducted as construction progresses. These meetings will cover coordination, deficiency resolution, and planning issues with particular Subs. The A/E will plan these meetings and will minimize unnecessary time being spent by Subs. For large projects, these meetings may be held monthly, until the final three months of construction when they may be held as frequently as one per week.

3.2 Reporting

A. The A/E will provide regular reports to the CM or PM, depending on the management structure, with increasing frequency as construction and commissioning progresses. Standard forms are provided and referenced in the *Commissioning Plan.*

B. The A/E will regularly communicate with all members of the commissioning team, keeping them apprised of commissioning progress and scheduling changes through memos, progress reports, etc.

C. Testing or review approvals and nonconformance and deficiency reports are made regularly, with the review and testing as described in later sections.

D. A final summary report (about four to six pages, not including backup documentation) by the A/E will be provided to the CM or PM, focusing on evaluating commissioning process issues and identifying areas where the process could be improved. All acquired documentation, logs, minutes, reports, deficiency lists, communications, findings, unresolved issues, etc., will be compiled in appendices and provided with the summary report. Prefunctional checklists, functional tests, and monitoring reports will not be part of the final report but will be stored in the Commissioning Record in the O&M manuals.

3.3 Submittals

A. The A/E will provide appropriate contractors with a specific request for the type of submittal needed for the commissioning work. These requests will be integrated into the normal submittal process and protocol of the construction team. At minimum, the request will include the manufacturer and model number, the manufacturer's printed installation and detailed startup procedures, full sequences of operation, O&M data, performance data, any performance test procedures, control drawings, and details of owner-contracted tests. In addition, the installation and checkout materials that are actually shipped inside the equipment and the actual field checkout sheet forms to be used by the factory or field technicians shall be submitted. All documentation requested will be included by the Subs in their O&M manual contributions.

B. The A/E will review and approve submittals related to the commissioned equipment for conformance to the contract documents as it relates to the commissioning process, to the functional performance of the equipment, and adequacy for developing test procedures. This review is intended primarily to aid in the development of functional testing procedures and only secondarily to verify compliance with equipment specifications. The A/E will notify the CM or PM, as requested, of items missing or areas that are not in conformance with contract documents and that require resubmission.

C. The A/E may request additional design narrative from the controls contractor, depending on the completeness of the design intent documentation and sequences provided with the specifications.

D. These submittals to the A/E do not constitute compliance for O&M manual documentation. The O&M manuals are the responsibility of the Contractor, though the A/E will review and approve them.

These guide specifications presume that design documentation and sequences of operation were carefully and completely prepared prior to bid documents being issued. If this was not the case, include language that requires the A/E and Controls Contractor to develop the design narrative and operating parameters according to the *design phase Commissioning Plan* and the format provided in Appendix 1 of that plan.

3.4 Startup, Prefunctional Checklists, and Initial Checkout

A. The following procedures apply to all equipment to be commissioned according to Section 1.7, "Systems to be Commissioned." Some systems that are not composed so much of actual dynamic machinery, e.g., electrical system power quality, may have very simplified PCs and startup.

B. **General.** Prefunctional checklists are important to ensure that the equipment and systems are hooked up and operational. It ensures that functional performance testing (in-depth system checkout) may proceed without unnecessary delays. Each piece of equipment receives full prefunctional checkout. No sampling strategies are used. The prefunctional testing for a given system must be successfully completed prior to formal functional performance testing of equipment or subsystems of the given system.

C. **Startup and Initial Checkout Plan.** The A/E shall assist the commissioning team members responsible for startup of any equipment in developing detailed startup plans. The primary role of the A/E in this process is to ensure that there is written documentation that each of the manufacturer-recommended procedures has been completed. Parties responsible for prefunctional checklists and startup are identified in the commissioning scoping meeting and in the checklist forms. Parties responsible for executing functional performance tests are identified in the testing requirements in Sections 15997, 16997, and [list other sections where tests requirements are found].

   1. The A/E adapts, if necessary, the representative prefunctional checklists and procedures from Section 15998. These checklists indicate required procedures to be executed as part of startup and initial checkout of the systems and the party responsible for their execution.

   2. These checklists and tests are provided by the A/E to the Contractor. The Contractor determines which trade is responsible for executing and documenting each of the line item tasks and notes that trade on the form. Most forms will have more than one trade responsible for their execution.

   3. The subcontractor responsible for the purchase of the equipment develops the full startup plan by combining (or adding to) the A/E's checklists with the manufacturer's detailed startup and checkout procedures from the O&M manual and the normally used field checkout sheets. The plan will include checklists and procedures with specific boxes or lines for recording and documenting the checking and inspections of each procedure and a summary statement with a signature block at the end of the plan.

The full startup plan could consist of something as simple as:

    a.   The prefunctional checklists.

    b.   The manufacturer's standard written startup procedures copied from the installation manuals with check boxes by each procedure and a signature block added by hand at the end.

    c.   The manufacturer's normally used field checkout sheets.

4.   The subcontractor submits the full startup plan to the A/E for review and approval.

5.   The A/E reviews and approves the procedures and the format for documenting them, noting any procedures that need to be added.

6.   The full startup procedures and the approval form may be provided to the CM for review and approval, depending on management protocol.

An alternative to the process for developing the startup plan given in parts 3 to 5 above consists of the A/E doing more of the work as described below.

OPTION FOR C. 3 to 5 ABOVE:

    a.   The A/E (instead of the contractor) copies the manufacturer's startup and initial checkout procedures from O&M submittals.

    b.   The A/E marks the applicable areas in the procedures and makes initial and date lines at each procedure or section.

    c.   The A/E transmits these procedures and the original prefunctional checklist procedures (see 1 above) to the contractor as the startup and initial checkout plan.

D. Sensor and Actuator Calibration.

All field-installed temperature, relative humidity, CO, CO_2, and pressure sensors and gages and all actuators (dampers and valves) on all equipment shall be calibrated using the methods described below. Alternative methods may be used, if approved by the Owner beforehand. All test instruments shall have had a certified calibration within the last 12 months. Sensors installed *in* the unit at the factory with calibration certification provided need not be field calibrated.

All procedures used shall be fully documented on the prefunctional checklists or other suitable forms, clearly referencing the procedures followed and written documentation of initial, intermediate, and final results.

Sensor Calibration Methods

All Sensors. Verify that all sensor locations are appropriate and away from causes of erratic operation. Verify that sensors with shielded cable are grounded only at one end. For sensor pairs that are used to determine a temperature or pressure difference, make sure they are reading within 0.2°F of each other for temperature and within a tolerance equal to 2% of the reading, of each other, for pressure. Tolerances for critical applications may be tighter.

Sensors without Transmitters. *Standard Application.* Make a reading with a calibrated test instrument within 6 inches of the site sensor. Verify that the sensor reading (via the permanent thermostat, gage, or building automation system [BAS]) is within the tolerances of the instrument-measured value in the table below. If not, install offset in BAS, calibrate or replace sensor.

Sensors With Transmitters. *Standard Application.* Disconnect sensor. Connect a signal generator in place of sensor. Connect ammeter in series between transmitter and BAS control panel. Using manufacturer's resistance-temperature data, simulate minimum desired temperature. Adjust transmitter potentiometer zero until 4 mA is read by the ammeter. Repeat for the maximum temperature matching 20 mA to the potentiometer span or maximum and verify at the BAS. Record all values and recalibrate controller as necessary to conform with specified control ramps, reset schedules, proportional relationship, reset relationship, and P/I reaction. Reconnect sensor. Make a reading with a calibrated test instrument within 6 inches of the site sensor. Verify that the sensor reading (via the permanent thermostat, gage, or building automation system) is within the tolerances of the instrument-measured value in the table below. If not, replace sensor and repeat. For pressure sensors, perform a similar process with a suitable signal generator.

Critical Applications. For critical applications (process, manufacturing, etc.) more rigorous calibration techniques may be required for selected sensors. Describe any such methods used on an attached sheet.

Tolerances, Standard Applications

Sensor	Required Tolerance (±)	Sensor	Required Tolerance (±)
Cooling coil, chilled and condenser water temps.	0.4°F	Flow rates, water Relative humidity	4% of design 4% of design
AHU wet-bulb or dew-point	2.0°F	Combustion flue temps.	5.0°F
Hot water coil and boiler water temp.	1.5°F	Oxygen or CO_2 monitor	0.1 % pts
Outside air, space air, duct air temps.	0.4°F	CO monitor	0.01 % pts
Watt-hour, voltage, and amperage	1% of design	Natural gas and oil flow rate	1% of design
Pressures, air, water, and gas	3% of design	Steam flow rate	3% of design
Flow rates, air	10% of design	Barometric pressure	0.1 in. of Hg

Valve and Damper Stroke Setup and Check

EMS Readout. For all valve and damper actuator positions checked, verify the actual position against the BAS readout. Set pumps or fans to normal operating mode. Command valve or damper closed, visually verify that valve or damper is closed and adjust output zero signal as required. Command valve or damper open, verify position is full open and adjust output signal as required. Command valve or damper to a few intermediate positions. If actual valve or damper position doesn't reasonably correspond, replace actuator or add pilot positioner (for pneumatics).

Closure for Heating Coil Valves (NO). Set heating setpoint 20°F above room temperature. Observe valve open. Remove control air or power from the valve and verify that the valve stem and actuator position do not change. Restore to normal. Set heating setpoint to 20°F below room temperature. Observe the valve close. For pneumatics, by override in the EMS, increase pressure to valve by 3 psi (do not exceed actuator pressure rating) and verify valve stem and actuator position does not change. Restore to normal.

Closure for Cooling Coil Valves (NC). Set cooling setpoint 20°F above room temperature. Observe the valve close. Remove control air or power from the valve and verify that the valve stem and actuator position do not change. Restore to normal. Set cooling setpoint to 20°F below room temperature. Observe valve open. For pneumatics, by override in the EMS, increase pressure to valve by 3 psi (do not exceed actuator pressure rating) and verify valve stem and actuator position does not change. Restore to normal.

E. Execution of Prefunctional Checklists and Startup.

1. Four weeks prior to startup, the Subs and vendors schedule startup and checkout with the CM, GC, and A/E. The performance of the prefunctional checklists, startup, and checkout are directed and executed by the Sub or vendor. When checking off prefunctional checklists, signatures may be required of other Subs for verification of completion of their work.

2. The A/E shall observe, at minimum, the procedures for each piece of primary equipment.

3. For lower-level components of equipment, (e.g., VAV boxes, sensors, controllers), the A/E shall observe a sampling of the prefunctional and startup procedures. The sampling procedures are identified in the commissioning plan.

4. The Subs and vendors shall execute startup and provide the A/E with a signed and dated copy of the completed startup and prefunctional tests and checklists.

5. Only individuals who have *direct* knowledge and witnessed that a line item task on the prefunctional checklist was actually performed shall initial or check that item off. It is not acceptable for witnessing supervisors to fill out these forms.

F. **Deficiencies, Nonconformance, and Approval in Checklists and Startup.**

1.   The Subs shall clearly list any outstanding items of the initial startup and prefunctional procedures that were not completed successfully at the bottom of the procedures form or on an attached sheet. The procedures form and any outstanding deficiencies are provided to the A/E within two days of test completion.

2.   The A/E reviews the report and submits either a noncompliance report or an approval form to the Sub or CM. The A/E shall work with the Subs and vendors to correct and retest deficiencies or uncompleted items. The A/E will involve the CM and others as necessary. The installing Subs or vendors shall correct all areas that are deficient or incomplete in the checklists and tests in a timely manner and shall notify the A/E as soon as outstanding items have been corrected and resubmit an updated startup report and a statement of correction on the original noncompliance report. When satisfactorily completed, the A/E recommends approval of the execution of the checklists and startup of each system to the CM using a standard form.

3.   Items left incomplete, which later cause deficiencies or delays during functional testing, may result in backcharges to the responsible party. Refer to Part 3.7, "Documentation, Nonconformance, and Approval of Tests," herein for details.

3.5 Phased Commissioning

A.   The project ___will require, ___will *not* require startup and initial checkout to be executed in phases. This phasing will be planned and scheduled in a coordination meeting of the A/E, CM, mechanical, TAB and controls, and the GC. Results will be added to the master and commissioning schedule.

3.6 Functional Performance Testing

A.   This subsection applies to all commissioning functional testing for all divisions.

B.   The general list of equipment to be commissioned is found in Section 17100, Part 1.4. The specific equipment and modes to be tested are found in Sections 15997, 16997, and [list other sections where tests requirements are found]   .

C.   The parties responsible to execute each test are listed with each test in Sections 15997, 16997, and [list other sections where tests requirements are found].

D.   **Objectives and Scope.** The objective of functional performance testing is to demonstrate that each system is operating according to the documented design intent and contract documents. Functional testing facilitates bringing the systems from a state of substantial completion to full dynamic operation. Additionally, during the testing process, areas of deficient performance are identified and corrected, improving the operation and functioning of the systems.

In general, each system should be operated through all modes of operation (seasonal, occupied, unoccupied, warmup, cooldown, part and full load) where there is a specified system response. Verifying each sequence in the sequences of operation is required. Proper responses to such modes and conditions as power failure, freeze condition, low oil pressure, no flow, equipment failure, etc., shall also be tested. Specific modes required in this project are given in Sections 15997, 16997, and [list other sections where tests requirements are found].

E. **Development of Test Procedures.** Before test procedures are written, the A/E shall obtain all requested documentation and a current list of change orders affecting equipment or systems, including an updated points list, program code, control sequences, and parameters. Using the testing parameters and requirements in Sections 15997, 16997, and [list other sections where tests requirements are found] the A/E shall develop specific test procedures and forms to verify and document proper operation of each piece of equipment and system. Each Sub or vendor responsible for executing a test shall provide limited assistance to the A/E in developing the procedures review (answering questions about equipment, operation, sequences, etc.). Prior to execution, the A/E shall provide a copy of the test procedures to the Sub(s) who shall review the tests for feasibility, safety, equipment, and warranty protection.

The A/E shall any review owner-contracted, factory testing or required owner acceptance tests that the A/E is not responsible to oversee, including documentation format, and shall determine what further testing or format changes may be required to comply with the *specifications*. Redundancy of testing shall be minimized.

The purpose of any given specific test is to verify and document compliance with the stated criteria of acceptance given on the test form.

Representative test formats and examples (not designed for this facility) are found in the appendices to Divisions 15 and 16. The test procedure forms developed by the A/E shall include (but not be limited to) the following information:

1. System and equipment or component name(s).
2. Equipment location and ID number.
3. Unique test ID number and reference to unique prefunctional checklist and startup documentation ID numbers for the piece of equipment.
4. Date.

5. Project name.

6. Participating parties.

7. A copy of the specification section describing the test requirements.

8. A copy of the specific sequence of operations or other specified parameters being verified.

9. Formulas used in any calculations.

10. Required pre-test field measurements.

11. Instructions for setting up the test.

12. Special cautions, alarm limits, etc.

13. Specific step-by-step procedures to execute the test, in a clear, sequential and repeatable format.

14. Acceptance criteria of proper performance with a Yes/No check box to allow for clearly marking whether or not proper performance of each part of the test was achieved.

15. A section for comments.

16. Signatures and date block for the A/E.

F. **Test Methods.**

1. Functional performance testing and verification may be achieved by manual testing (persons manipulate the equipment and observe performance) or by monitoring the performance and analyzing the results using the control system's trend log capabilities or by stand-alone dataloggers. Sections 15997, 16997, and [list other sections where tests requirements are found] specify which methods shall be used for each test. The A/E may substitute specified methods or require an additional method to be executed, other than what was specified, with the approval of the CM. This may require a change order and adjustment in charge to the owner. The A/E will determine which method is most appropriate for tests that do not have a method specified.

2. **Simulated Conditions.** Simulating conditions (not by an overwritten value) shall be allowed, though timing the testing to experience actual conditions is encouraged wherever practical.

3. **Overwritten Values.** Overwriting sensor values to simulate a condition, such as overwriting the outside air temperature reading in a control system to be something other than it really is, shall be allowed, but shall be used with caution and avoided when possible. Such testing methods often can only test a part of a system, as the interactions and responses of other systems will be erroneous or not applicable. Simulating a condition is preferable (e.g., for the above case, by heating the outside air sensor with a hair blower rather than overwriting the value or by altering the appropriate setpoint to see the desired response). Before simulating conditions or overwriting values, sensors, transducers, and devices shall have been calibrated.

4. **Simulated Signals.** Using a signal generator, which creates a simulated signal to test and calibrate transducers and DDC constants, is generally recommended over using the sensor to act as the signal generator via simulated conditions or overwritten values.

5. **Altering Setpoints.** Rather than overwriting sensor values, and when simulating conditions is difficult, altering setpoints to test a sequence is acceptable. For example, to see the AC compressor lockout work at an outside air temperature below 55°F, when the outside air temperature is above 55°F, temporarily change the lockout setpoint to be 2°F above the current outside air temperature.

6. **Indirect Indicators.** Relying on indirect indicators for responses or performance shall be allowed only after visually and directly verifying and documenting, over the range of the tested parameters, that the indirect readings through the control system represent actual conditions and responses. Much of this verification is completed during prefunctional testing.

7. **Setup.** Each function and test shall be performed under conditions that simulate actual conditions as close as is practically possible. The Sub executing the test shall provide all necessary materials, system modifications, etc., to produce the necessary flows, pressures, temperatures, etc., necessary to execute the test according to the specified conditions. At completion of the test, the Sub shall return all affected building equipment and systems, due to these temporary modifications, to their pre-test condition.

8. **Sampling.** Each heat pump and its connected ancillary equipment (two-way valves, balance valves, circulator pumps, thermostats, etc.) shall be individually tested, including tests for leaks, sequence of operations, and performance.

G. **Coordination and Scheduling.** The Subs shall provide sufficient notice to the A/E regarding their completion schedule for the prefunctional checklists and startup of all equipment and systems. The A/E will schedule functional tests through the CM, GC, and affected Subs. The A/E shall direct, witness, and document the functional testing of all equipment and systems. The Subs shall execute the tests.

In general, functional testing is conducted after prefunctional testing and startup has been satisfactorily completed. The control system is sufficiently tested and approved by the A/E before it is used for TAB or to verify performance of other components or systems. The air balancing and water balancing is completed and debugged before functional testing of air-related or water-related equipment or systems. Testing proceeds from components to subsystems to systems. When the proper performance of all interacting individual systems has been achieved, the interface or coordinated responses between systems is checked.

H. **Test Equipment.** Refer to Section 17100, Part 2, for test equipment requirements.

I. **Problem Solving.** The A/E will recommend solutions to problems found; however, the burden of responsibility to solve, correct and retest problems is with the GC and Subs.

3.7 Documentation, Nonconformance, and Approval of Tests

A. **Documentation.** The A/E shall witness and document the results of all functional performance tests using the specific procedural forms developed for that purpose. Prior to testing, these forms are provided to the CM for review and approval and to the Subs for review. The A/E will include the filled-out forms in the O&M manuals.

B. **Nonconformance**.

1. The A/E will record the results of the functional test on the procedure or test form. All deficiencies or nonconformance issues shall be noted and reported to the CM on a standard noncompliance form.

2. Corrections of minor deficiencies identified may be made during the tests at the discretion of the A/E. In such cases the deficiency and resolution will be documented on the procedure form.

3. Every effort will be made to expedite the testing process and minimize unnecessary delays, while not compromising the integrity of the procedures. However, the A/E will not be pressured into overlooking deficient work or loosening acceptance criteria to satisfy scheduling or cost issues, unless there is an overriding reason to do so at the request of the CM.

4. As tests progress and a deficiency is identified, the A/E discusses the issue with the executing contractor.

    a. When there is no dispute on the deficiency and the Sub accepts responsibility to correct it:

       1) The A/E documents the deficiency and the Sub's response and intentions and they go on to another test or sequence. After the day's work, the A/E submits the noncompliance reports to the CM for signature, if required. A copy is provided to the Sub and A/E. The Sub corrects the deficiency, signs the statement of correction at the bottom of the noncompliance form certifying that the equipment is ready to be retested, and sends it back to the A/E.

       2) The A/E reschedules the test and the test is repeated.

    b. If there is a dispute about a deficiency, regarding whether it is a deficiency or who is responsible:

1) The deficiency shall be documented on the noncompliance form with the Sub's response and a copy given to the CM and to the Sub representative assumed to be responsible.

2) Resolutions are made at the lowest management level possible. Other parties are brought into the discussions as needed. Final interpretive authority is with the A/E. Final acceptance authority is with the Project Manager.

3) The A/E documents the resolution process.

4) Once the interpretation and resolution have been decided, the appropriate party corrects the deficiency, signs the statement of correction on the noncompliance form, and provides it to the A/E. The A/E reschedules the test and the test is repeated until satisfactory performance is achieved.

5. Cost of Retesting.

a. The cost for the *Sub* to retest a prefunctional or functional test, if they are responsible for the deficiency, shall be theirs. If they are not responsible, any cost recovery for retesting costs shall be negotiated with the GC.

b. For a deficiency identified, not related to any prefunctional checklist or startup fault, the following shall apply: The A/E and CM will direct the retesting of the equipment once at no "charge" to the GC for their time. However, the A/E's and CM's time for a second retest will be charged to the GC, who may choose to recover costs from the responsible Sub.

c. The time for the A/E and CM to direct any retesting required because a specific *prefunctional* checklist or startup test item, reported to have been successfully completed but determined during functional testing to be faulty, will be backcharged to the GC, who may choose to recover costs from the party responsible for executing the faulty prefunctional test.

d. Refer to the sampling section of Section 17100, Part 3.6 for requirements for testing and retesting identical equipment.

6. The Contractor shall respond in writing to the A/E and CM at least as often as commissioning meetings are being scheduled concerning the status of each apparent outstanding discrepancy identified during commissioning. Discussion shall cover explanations of any disagreements and proposals for their resolution.

7. The A/E retains the original nonconformance forms until the end of the project.

8. Any required retesting by any contractor shall not be considered a justified reason for a claim of delay or for a time extension by the prime contractor.

C. **Failure Due to Manufacturer Defect.** If 10%, or three, whichever is greater, of identical pieces (size alone does not constitute a difference) of equipment fail to perform to the contract documents (mechanically or substantively) due to manufacturing defect, not allowing it to meet its submitted performance specifications, all identical units may be considered unacceptable by the CM or PM. In such case, the contractor shall provide the owner with the following:

   a. Within one week of notification from the CM or PM, the contractor or manufacturer's representative shall examine all other identical units making a record of the findings. The findings shall be provided to the CM or PM within two weeks of the original notice.

   b. Within two weeks of the original notification, the contractor or manufacturer shall provide a signed and dated, written explanation of the problem, cause of failures, etc., and all proposed solutions, which shall include full equipment submittals. The proposed solutions, shall not significantly exceed the specification requirements of the original installation.

   c. The CM or PM will determine whether a replacement of all identical units or a repair is acceptable.

   d. Two examples of the proposed solution will be installed by the contractor and the CM will be allowed to test the installations for up to one week, upon which the CM or PM will decide whether to accept the solution.

   e. Upon acceptance, the contractor and/or manufacturer shall replace or repair all identical items at their expense and extend the warranty accordingly, if the original equipment warranty had begun. The replacement/repair work shall proceed with reasonable speed beginning within one week from when parts can be obtained.

D. **Approval.** The A/E notes each satisfactorily demonstrated function on the test form. Formal approval of the functional test is made later after review by the A/E and by the CM, if necessary. The A/E recommends acceptance of each test to the CM using a standard form. The CM gives final approval on each test using the same form, providing a signed copy to the A/E and the contractor.

3.8 Operation and Maintenance Manuals

> The following O&M documentation requirements assume that the general contractor is compiling the O&M manuals, with all Subs compiling their own sections, including some submissions for the A/E.
>
> These requirements may need to be merged and edited to follow the protocols and scope of the current agency or project. However, the comprehensiveness and accessibility described herein shall be maintained.

A. Standard O&M Manuals.

1. The specific content and format requirements for the standard O&M manuals are detailed in Section 01730. Special requirements for the controls contractor and TAB contractor are found in Section 15995, Part 3.6.

2. **A/E Contribution.** The A/E will include in the beginning of the O&M manuals a separate section describing the systems including:

   a. The design intent narrative prepared by the A/E and provided as part of the bid documents, updated to as-built status by the A/E.

   b. Simplified, professionally drawn, single-line system diagrams on 8 ½ in. × 11 in. or 11 in. × 17 in. sheets. These shall include chillers, water system, condenser water system, heating system, supply air systems, exhaust systems, environmental heat exchanger (ground-coupled, surface-water, groundwater, or other) and _____. These shall show major pieces of equipment such as pumps, boilers, control valves, expansion tanks, coils, service valves, etc.

3. **Commissioning Review and Approval.** Prior to substantial completion, the A/E shall review the O&M manuals, documentation, and redline as-builds _for systems that were commissioned_ and [list other systems documentation that the CA should review] to verify compliance with the _specifications._ The A/E will communicate deficiencies in the manuals to the CM or PM, as requested. Upon a successful review of the corrections, the A/E recommends approval and acceptance of these sections of the O&M manuals to the CM or PM. The A/E also reviews each equipment warranty and verifies that all requirements to keep the warranty valid are clearly stated. This work is part of the A/E's review of the O&M manuals according to the A/E's contract.

B. Commissioning Record in O&M Manuals.

1. The A/E is responsible for compiling, organizing, and indexing the following commissioning data by equipment into labeled, indexed and tabbed, three-ring binders and delivering it to the GC, to be included with the O&M manuals. Three copies of the manuals will be provided. The format of the manuals shall be:

    Tab I-1   Commissioning Plan

    Tab I-2   Final Commissioning Report (see (B.2) below)

    Tab 01   System Type 1   (chiller system, packaged unit, boiler system, etc.)

    > *Sub-Tab A*   Design narrative and criteria, sequences, approvals for Equipment 1
    >
    > *Sub-Tab B*   Startup plan and report, approvals, corrections, blank prefunctional checklists
    >
    > *Colored Separator Sheets*—for each equipment type (fans, pumps, chiller, etc.)
    >
    > *Sub-Tab C*   Functional tests (completed), trending and analysis, approvals and corrections, training plan, record and approvals, blank functional test forms, and a recommended recommissioning schedule.

    Tab 02   System Type 2......repeat in accordance with System 1

2. **Final Report Details.** The final commissioning report shall include an executive summary, list of participants and roles, brief building description, overview of commissioning and testing scope, and a general description of testing and verification methods. For each piece of commissioned equipment, the report should contain the disposition of the commissioning authority regarding the adequacy of the equipment, documentation, and training meeting the contract documents in the following areas: 1) equipment meeting the equipment specifications, 2) equipment installation, 3) functional performance and efficiency, 4) equipment documentation and design intent, and 5) operator training. All outstanding noncompliance items shall be specifically listed. Recommendations for improvement to equipment or operations, future actions, commissioning process changes, etc., shall also be listed. Each noncompliance issue shall be referenced to the specific functional test, inspection, trend log, etc., where the deficiency is documented. The functional performance and efficiency section for each piece of equipment shall include a brief description of the verification method used (manual testing, BAS trend logs, data loggers, etc.) and include observations and conclusions from the testing.

3. Other documentation will be retained by the A/E.

3.9 Training of Owner Personnel

A. The GC shall be responsible for training coordination and scheduling and ultimately for ensuring that training is completed.

B. The A/E shall be responsible for overseeing and approving the content and adequacy of the training of owner personnel for commissioned equipment.

    1. The A/E shall interview the facility manager and lead engineer to determine the special needs and areas where training will be most valuable. The owner and A/E shall decide how rigorous the training should be for each piece of commissioned equipment. The A/E shall communicate the results to the Subs and vendors who have training responsibilities.

    2. In addition to these general requirements, the specific training requirements of Owner personnel by Subs and vendors is specified in Division 15 and 16 and [list other sections where training requirements are found].

    3. Each Sub and vendor responsible for training will submit a written training plan to the A/E for review and approval prior to training. The plan will cover the following elements:

        a. Equipment (included in training)

        b. Intended audience

        c. Location of training

        d. Objectives

        e. Subjects covered (description, duration of discussion, special methods, etc.)

        f. Duration of training on each subject

        g. Instructor for each subject

        h. Methods (classroom lecture, video, site walk-through, actual operational demonstrations, written handouts, etc.)

        i. Instructor and qualifications

    4. For the primary HVAC equipment, the controls contractor shall provide a short discussion of the control of the equipment during the mechanical or electrical training conducted by others.

    5. The A/E develops an overall training plan and coordinates and schedules, with the CM and GC, the overall training for the commissioned systems. The A/E develops criteria for determining that the training was satisfactorily completed, including attending some of the training, etc. The A/E recommends approval of the training

to the CM using a standard form. The CM also signs the approval form.

6. At one of the training sessions, the A/E presents a _____ hour presentation discussing the use of the blank functional test forms for recommissioning equipment.

7. Videotaping of the training sessions will be provided by ___ the owner, ___ the A/E with tapes cataloged by ___ the owner, ___ the A/E and added to the O&M manuals.

8. The mechanical design engineer shall, at the first training session, present the overall system design concept and the design concept of each equipment section. This presentation shall be _____ hours in length and include a review of all systems using the simplified system schematics (one-line drawings) including chilled water systems, condenser water or heat rejection systems, heating systems, fuel oil and gas supply systems, supply air systems, exhaust system, and outside air strategies.

3.10 Deferred Testing

A. **Unforeseen Deferred Tests.** If any check or test cannot be completed due to the building structure, required occupancy condition, or other deficiency, execution of checklists and functional testing may be delayed upon approval of the PM. These tests will be conducted in the same manner as the seasonal tests as soon as possible. Services of necessary parties will be negotiated.

B. **Seasonal Testing.** During the warranty period, seasonal testing (tests delayed until weather conditions are closer to the system's design) specified in Section 15997 shall be completed as part of this contract. The A/E shall coordinate this activity. Tests will be executed, documented, and deficiencies corrected by the appropriate Subs, with facilities staff and the A/E witnessing. Any final adjustments to the O&M manuals and as-builts due to the testing will be made.

3.11 Written Work Products

A. The commissioning process generates a number of written work products described in various parts of the *specifications*. The *commissioning plan—construction phase,* lists all the formal written work products, describes briefly their contents, who is responsible for creating them, their due dates, who receives and approves them, and the location of the specification to create them. In summary, the written products are as follows:

Product	Developed By
1. Final commissioning plan	A/E
2. Meeting minutes	A/E
3. Commissioning schedules	A/E with GC and CM
4. Equipment documentation submittals	Subs

5. Sequence clarifications	Subs and A/E as needed
5. Prefunctional checklists	A/E (already in Specs)
6. Startup and initial checkout plan	Subs and A/E (compilation of existing documents)
7. Startup and initial checkout forms filled out	Subs
8. Final TAB report	TAB
9. Issues log (deficiencies)	A/E
10. Commissioning Progress Record	A/E
11. Deficiency reports	A/E
12. Functional test forms	A/E
13. Filled-out functional tests	A/E
14. O&M manuals	Subs
15. Commissioning record book	A/E
16. Overall training plan	A/E and CM
17. Specific training agendas	Subs
18. Final commissioning report	A/E
19. Miscelaneous approvals	A/E

APPENDIX A-2—TOPICS FOR COMMISSIONING

CSI Sections treated in ORNL/FEMP Commissioning Guide.[1]

Number	Title	Number of Pages
00050	Applicable Sections For Geothermal Heat Pump Systems	8
01000	Site-Specific Specification	1
02050	Demolition	3
02110	Site Clearing	2
02200	Earthwork	6
02225	Trenching	5
02270	Erosion Control	4
02276	Geotextile Fabric	4
02300	Boring and Jacking	4
02505	Mineral Aggregate Base Course	3
02510	Asphalitic Concrete Paving	4
02520	Screen and/or Gravel Pack—Unconsolidated Aquifer	29
02525	Open Hole Completion—Consolidated Formation	25
02936	Seeding	4
02938	Sodding	3
03600	Thermal-Enhanced Bentonite Grout	4
15050	Piping Systems	10
15052	Brazing	2
15071	Underground Protective Coating	3
15072	Cleaning	5
15073	Pressure/Leak Testing	10
15074	Identification and Labeling	4
15100	Valves	7
15101	Steam and Condensate (0-150 Psig)	6
15106	Chilled Water, Cooling Water, Process Water, and Heat Pump Water	5
15110	Geothermal Heat Pump and Loop Piping Systems (High Density Polyethylene)	7

---

1. Thomas, W., and M. Madgett. 2000. Generic Guide Specifications for Geothermal Heat Pump System Installation. ORNL/TM-2000/132, prepared under U.S. DOE Contract DE-AC05-00OR22725.

Number	Title	Number of Pages
15112	Instrument Air	3
15125	Fluorinated Hydrocarbon Refrigerants	5
15135	Thermometers and Gauges	5
15260	Piping Insulation	8
15262	Fibrous Glass Insulation	4
15265	Elastomeric Rubber Insulation	4
15270	Aluminum Jacketing	2
15292	Thermal Insulation for Ductwork—Fibrous Glass Panels w/AL Facing	3
15293	Thermal Insulation for Ductwork—Fibrous Glass Blanket w/AL Facing	3
15501	Heating, Ventilating, and Air-Conditioning Systems—Installation And Equipment	5
15515	Exterior Ground Loop Heat Exchangers	11
15540	Water Circulating Pumps For HVAC	6
15545	Submersible Well Pumps	3
15550	Open Lineshaft, Water-Lubricated Turbine Pump	3
15555	Oil-Lubricated, Enclosed Lineshaft Turbine Pump	4
15786	Water-Source Heat Pumps	12
15891	Ductwork—Galvanized Steel, Low Velocity and Low Pressure	5
15902	Ductwork—Insulated and Non-insulated Flexible	3
15950	Testing, Adjusting, and Balancing	8
15951	Control Systems	4
15952	Copper Control Tubing w/ Compression Fittings	4
15954	Plastic Control Tubing—Pneumatic Controls	6
15955	Direct Digital Controls	17
15956	Control Panels and Instruments	6
15970	Plate Heat Exchangers	2
15972	Packaged Cooling Towers	4
15975	Field-Erected Cooling Towers	5
15980	Closed-Circuit Coolers	3
15995	Mechanical System Commissioning	15
16050	Basic Materials and Methods	2
16060	Electrical Demolition	3
16111	Conduit and Fittings	4
16120	Building Wire and Cable—600 Volts and Below	7

Number	Title	Number of Pages
16127	Splices and Terminations—600 Volts and Below	9
16131	Boxes	3
16141	Wiring Devices	4
16160	Equipment Cabinets and Enclosures	3
16191	Supporting Devices	2
16196	Electrical Identification	4
16441	Disconnect Switches	3
16451	Secondary Grounding	4
16471	Panelboards	4
16476	Enclosed Circuit Breakers	3
16481	Motor Control Centers	9
16482	Motor-operated Equipment	2
16483	Motor Control	6
16484	Adjustable Frequency Drive for A-C Induction Motors	6
16485	Contactors	3
16942	Field Components Installation	8
16943	Consoles, Panels, Cabinets, and Racks Installation	3
16960	Electrical Testing	3

Appendix B
Selected Troubleshooting Remedies

This appendix describes the procedures for corrective actions unique to geothermal systems. Items B.1 through B.3 are applicable to ground heat exchanger systems[1] and items B.4 through B.7 are applicable to groundwater (well) systems. The cross-reference to the "Troubleshooting Guide" (chapter 4) is given at the beginning of each section.

The information in this appendix is intended as a guide only, and it must not be used as a substitute for the manufacturer's recommended procedures or established safe practices in each specific trade. When testing and repairing GSHP systems, all applicable safety and environmental regulations must be followed.

B.1 FLUSHING GROUND LOOPS (SEE TABLE 4-3 t, x)

1. To ensure adequate flow rates for flushing, each individual loop, rather than the entire ground heat exchanger, should be flushed at any one time.
2. Access each loop at the isolation valves of the supply and return runouts.
3. Flush clean water at a minimum velocity of 2.5 ft/sec for 5 minutes in each direction.

B.2 LOCATING GROUND HEAT EXCHANGER LEAKS (SEE TABLE 4-3 g, p, aa)

Horizontal ground loops:
1. Determine the loop where the leak is located by looking for areas of saturated ground above the ground heat exchanger.
2. If the ground is saturated over the whole field, it may be possible to find the leak by injecting an inexpensive gas (such as nitrogen, carbon dioxide, or even air) and looking for bubbles at the top of the saturated ground. Of course, the system must be purged after introducing any gas and after the leak is fixed.
3. If these measures fail, there are commercial services that inject proprietary tracer gases. They may use highly sensitive atomic absorption instruments carried around the field to find high concentrations of the gas and, thus, the leak locations. This can be more sensitive than looking for bubbles.

---

1. Including closed-loop surface water heat pump systems.

Vertical ground heat exchangers:

Leaks in the vertical U-bends are very rare, particularly when the installer has carried out the required pressure tests before inserting the loop and after grouting. Unless the leak is in a run-out trench, it will be difficult to locate the ground saturation level, due to the depth of the tubes. Tracer gas will probably not be effective at these depths.

1. Check first for leaks in the run-outs by looking for saturated ground in the run-out trench locations. If no saturated area is found, proceed to steps 2 and 3.

2. Pressure test each run-out to determine the loop where the U-tube that leaks is located.

3. Pressure test each individual U-tube on this loop to determine the specific U-tube that leaks.

Although leaks have been found with these methods, the challenges of isolating loop leaks have led some designers to protect themselves with relatively inexpensive design precautions. These include limiting the number of boreholes per circuit and requiring enough total loop length that any single circuit can be abandoned without degrading performance below the specification. Another common precaution against loop inadequacies is to specify manifolds with unused "stubs," to allow connecting a new circuit or a fluid cooler if necessary.

B.3 REPAIRING HDPE GROUND HEAT EXCHANGER PIPING (SEE TABLE 4-3 g, l, m, p, z, aa, ak, am)

Permanent Repair

1. Locate cracked or damaged pipe section (see Section B.2)

2. Excavate around the pipe taking care not to cause further damage. Remove enough soil to allow access for a socket fusion machine.

3. Due to the flexibility of small diameter pipes, the defective area can be replaced with two flanged ends fused directly to each protruding end (see Figure B1a). Large diameter pipes will require a spool section to be added (see Figure B1b) since the pipe cannot be deflected to provide access for socket fusing. Remove the amount of pipe required to fit the scheme adopted. The length of pipe to be removed should be accurately determined so as to avoid unnecessary stresses in the pipe when the joint is bolted together.

4. Socket fuse the flanges to the pipe using a socket fusion machine. (For large diameter pipe prepare the spool assembly by butt fusing the flanges to a small section of pipe.)

5. Bolt the flanges together. Leak test.

6. Backfill the repair trench and compact the ground.

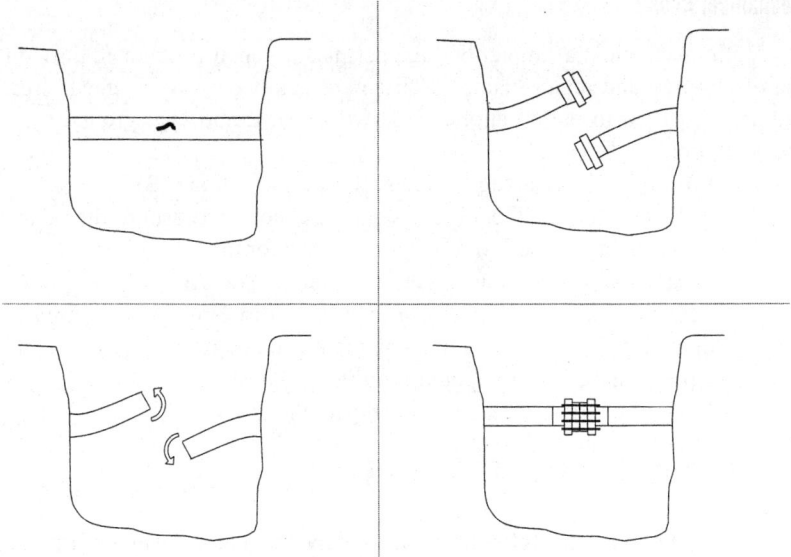

Figure B1a   Repairing small diameter HDPE piping.

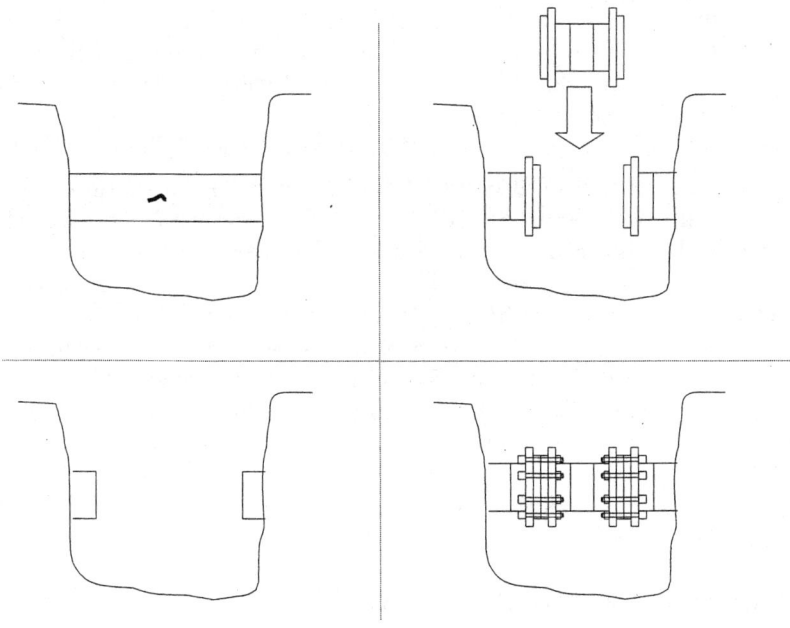

Figure B1b   Repairing large diameter HDPE piping.

Mechanical Repair

With this method a simpler but less permanent repair is obtained through the use of wrap-around repair clamps. These are equipped with an integral gasket and are best used in buried applications where the compacted ground prevents excessive pull out forces.

1. Locate crack or damaged pipe section (see Section B.2).
2. Excavate around the pipe taking care not to cause further damage. Remove enough soil to provide clearance for the clamp.
3. Clean pipe of all foreign material around the defect.
4. Select a clamp of a minimum length 1.25 to 2 times the nominal pipe diameter (or as recommended by the manufacturer).
5. Apply the clamp and tighten evenly. Leak test.
6. Backfill the repair trench and compact the ground.

B.4 WELL REDEVELOPMENT (SEE TABLE 4-4 a, e)

The type of well redevelopment will be dependent on the cause of the performance degradation. This should be determined in advance by chemical testing of the well and by video inspection of the well upon removal of the submersible pump. The following mechanical and chemical redevelopment methods should be attempted first before resorting to the more costly procedures.

1. *Clogging of the screen with fines*: use surging and swabbing to dislodge the fines.
2. *Calcareous incrustation*: treat the well with an acid solution approved for the purpose. This can be combined with wire brush and/or high pressure water jet cleaning.
3. *Iron bacteria*: treat the well with an approved acid solution followed by high-pressure water jet cleaning. For small diameter wells, a chlorine application can be used instead of the acid. No treatment will completely kill the bacteria in the well, so the treatment should be scheduled according to the needs of the particular well.
4. *Clogging of the screen with clay*: use an approved dispersing agent.
5. *Clogging of the gravel pack*: High-pressure water jet cleaning of the screen will also serve to redevelop the gravel pack behind the screen. Surging will also help to clear the gravel pack.

If these methods are ineffective the following more expensive steps can be taken.

1. Replace the entire screen.
2. Deepen the well.

B.5 REPLACING A SUBMERSIBLE PUMP (SEE TABLE 4-4 r, s, t)

1. Disconnect power to the submersible pump at the electrical box.
2. Disconnect wires to the pump at the well head.

3. Close the shutoff valve at the discharge elbow.
4. Open the air/vac valve to allow the water in the well discharge column to descend fully into the well.
5. Disconnect the pipe at the outlet flange.
6. Remove the well cover.
7. Place the crane hook around the lifting ring.
8. Raise the column with the crane until an entire column section is above the well mouth.
9. Support the column by the section beneath and remove the top section.
10. Raise the next section up with the crane until it is entirely above the well mouth.
11. Repeat steps 9 and 10 until the pump and screen are retrieved.
12. Replace the pump.
13. Clean the screen and inspect for damage or corrosion. Replace if necessary.
14. Inspect the wire condition and replace if necessary.
15. Reassemble by reversing the procedure.

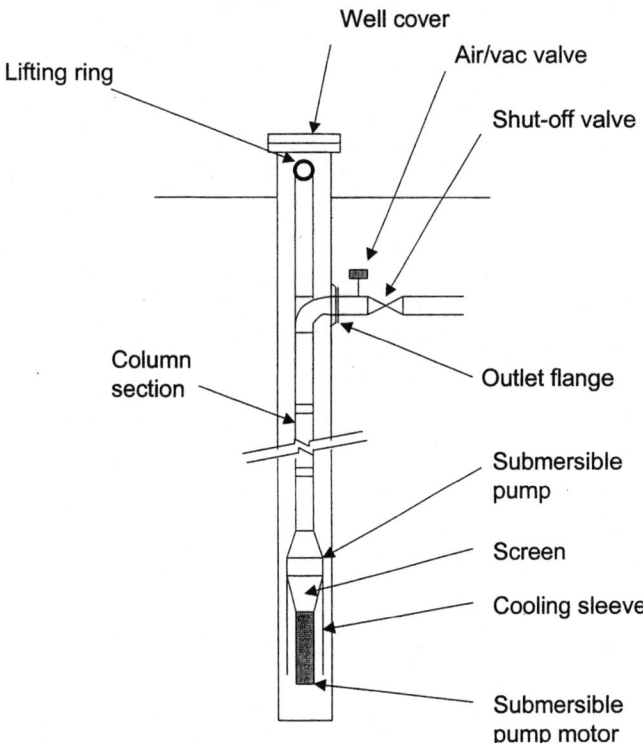

Figure B2 Typical submersible pump installation.

B.6 CLEANING A PLATE-FRAME HEAT EXCHANGER (SEE TABLE 4-4 b, c)

1. Remove heat exchanger protective shield (if fitted).
2. Unbolt and remove tightening rods.
3. Remove adjustable pressure plate.
4. Remove plates from carrying bar.
5. Remove gaskets from plates, taking care not to damage them.
6. Clean plates (do not use a metal brush unless necessary—if so, use a brush of the same type of metal as the plates to avoid surface damage).
7. Inspect plates for damage or cracks. Replace all damaged plates.
8. Inspect gaskets for damage or degradation. Replace where necessary.
9. Reassemble plates.
10. Install adjustable pressure plate.
11. Install tightening bars.
12. Torque tighten rod nuts following manufacturer's recommended tightening sequence and torque. Overtorquing can damage gaskets.
13. Order spare plates and gaskets to replace those used.

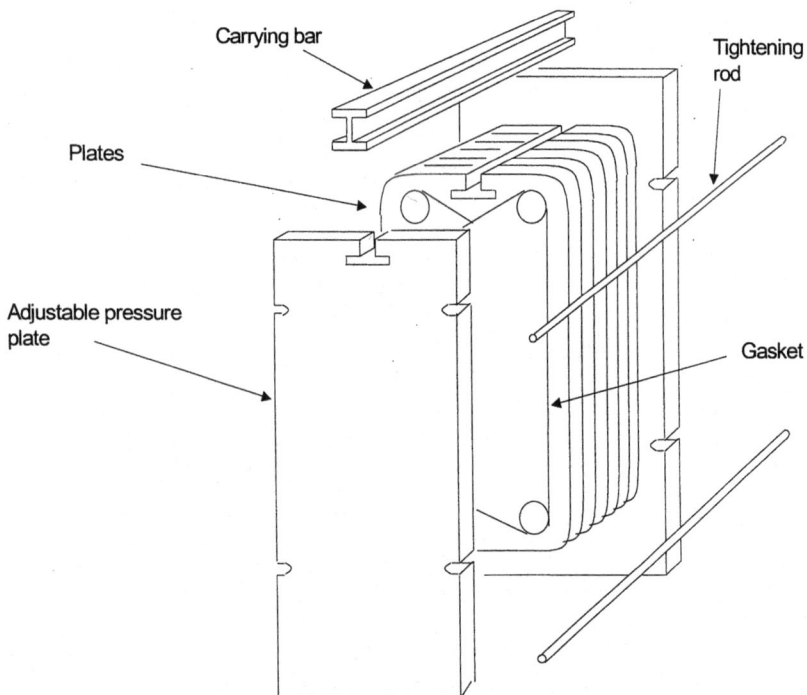

Figure B3 Exploded view of a plate-frame heat exchanger.

Table B-7
Controller Temperature Range for Dual Setpoint Control °C (°F)

Motor power		System Thermal Mass L/(kW of peak block load) or US gal/(ton of peak block load)						
kW	hp	2	4	6	8	10	12	14
Cooling Mode								
<3.7	<5	16 (28)	8 (14)	5 (9)	4 (7)	3.3 (6)	3 (5)	2 (4)
>3.7	>5	31 (56)	16 (28)	11 (19)	8 (14)	6 (11)	5 (9)	4 (8)
Heating Mode								
<3.7	<5	9 (16)	4 (8)	3 (5)	2 (4)	2 (3)	2 (3)	1 (1)
>3.7	>5	18 (32)	9 (16)	6 (11)	4 (8)	3 (6)	3 (5)	3 (5)

B.7 DETERMINING THE TEMPERATURE RANGE FOR A WELL PUMP SETPOINT CONTROL (DUAL) (SEE TABLE 4-4 o, r)

In order to reduce the pump and motor wear brought about from cycling on and off, an adequate setpoint range for the pump controller must be determined. This range must also not be too great since this will result in reduction in heat pump efficiency. A smaller range should be specified for the heating season than for the cooling season. Table B-7 (taken from Rafferty [2000]) gives recommended setpoint ranges based on the pump motor size and system thermal mass. Note that L/kW of block load is approximately equal to USgal/ton of block load. The large setpoint ranges when the thermal mass is less that 8 L/kW of peak block load indicate that for these cases some mass should be added to the loop. The size of storage tank required can be estimated from the table.

Where multiple wells are employed, staged groundwater pumping can be installed. This allows the controller temperature ranges to be reduced, lessening the importance of system thermal mass.

Appendix C
Prefunctional Checks and Tests for GSHPs

C.1 VERTICAL GROUND HEAT EXCHANGER

Piping for vertical heat exchanger installation should be delivered to the site sealed and under air pressure. Seals keep dirt out. Pressurization at the factory reveals defects: If cutting the pipe open does not release pressure, the pipe has pinhole or other defects. Before installation, each circuit in a vertical heat exchanger installation shall be filled with water (air tests are not acceptable) and pressure tested for leaks and integrity at a pressure of 100 psi (with due regard to the static head at the lowest point in the loop so as not to exceed the burst pressure rating of the pipe) for a period of four hours. The maximum acceptable drop in pressure during this period should be no more than 5 psi. Seal and insert the heat exchanger into the borehole and immediately tremi-grout. Seal by crimping/capping and fusing. Maintain the pressure for one additional hour after completion of the tremi-grouting.

Pressure testing. Larger systems should have four hydrostatic pressure tests. Two methods are available. The first is outlined in IGSHPA (1997, p. 4 (item 1E). It involves flushing the pipe of debris and purging all air. Then the pipe is put under the smaller of 150% of the pipe design or 300% of the system operating pressure for 30 minutes without leaks. Ideally, the pipe should be raised to test pressure and allowed to stand for two to three hours to allow for pipe expansion. After equilibrium is established, the final test pressure is recorded each half-hour. The other method has been developed by a pipe manufacturer.[a] This is a four-step process. After flushing and purging, the pipe is put under pressure. The pipe is held at pressure for 45 minutes, and then the pressure is rapidly brought down to 30 to 40 psi (2.1 to 2.8 kg/cm^2). The pressure is then measured at specific time intervals for the next 90 minutes. Air-free and leak-free pipe will show a distinct, monotonic, pressure recovery as the pipe recovers its dimension at the reduced pressure. The four tests should be done on the

- pipe, before insertion;
- pipe, after insertion in the ground, to ensure that the pipe was not cut during insertion;
- circuit (assembly of 6 to 14 U-bends on a single header) after assembly but before backfilling, to ensure that all joints have complete integrity and that there is no pipe damage;
- external system at the interior manifold, for system integrity.

The effort required for these tests is much less than the consequences if they are not done and a problem is discovered later.

---

a. Performance Pipe, Tech Note 35, 1999.

Assemble the circuit headers and runouts. Thermally fuse to pipe manufacturer's recommendation. Pressure test the circuit headers and runouts prior to assembling to the circuits. Once assembled, backfill the return and supply runouts in separate trenches or separated from one another in the same trench. Before backfilling the trenches, pressure test the system at 100 psi as described above. Ensure that air and debris are purged and flushed from the system using a portable, charging station.

On-site testing, verification, and reporting should be undertaken as described in the following:

1. All borehole locations and depths and heat exchanger lengths are as required by the design.
2. Verification of the type of grout, mixing procedures, and topping off of boreholes with grout is to be done on an ongoing basis during ground heat exchanger installation.
3. Oversee the pressure testing of the circuits, circuit headers, and main runouts as described earlier to ensure there are no leaks.
4. Oversee the balancing of the different runouts, if required.
5. Verify that the antifreeze solution and chemical inhibitor characteristics and concentration are as called for in the design.
6. Prepare a report for the owner on the above testing and verification and ensure that the "as-built" conditions are recorded on the design plans.

C.2 GROUNDWATER SYSTEMS

The purpose of flow testing (from the engineer's standpoint) is to provide the information necessary to proceed with design (information necessary to calculate well pump power at various flows). Long-term constant flow rates are rarely if ever done for GSHP systems. Step tests are the common method, usually of 4 to 12 hours' duration. Length of time at any particular flow is determined by the stablilization of water level, not time. These tests provide water level, flow, drawdown, specific capacity, temperature, and static water level data. Nearby wells are not typically monitored in step tests. These tests need not be monitored by a hydrogeologist—pump contracters are familiar with and capable of performing them. Experienced engineers are capable of interpreting the results to the extent necessary for GSHP design. Constant flow tests where nearby wells are monitored produce information on aquifer characteristics (transmissivity and storage coefficient). These tests should involve the services of a hydrogeologist.

Underground Piping. The piping that connects the well(s) to the mechanical room is buried below the frost line. Typically it is uninsulated and frequently polyethylene pressure pipe (Class 160) is used since it is impervious to both interior and exterior corrosion and provides for a low-cost installation. Plumbing for each well should be independent, but common trenches for multiple-well systems are recommended wherever practical. Electrical supply cables for the pumps may also use the same trenches, but these must also meet the applicable electrical code.

Pressure test the piping and confirm correct electrical function prior to back-filling the trenches. Carefully backfill the first 6 inches of dirt over each layer of pipe and cable, taking care to exclude sharp rocks or other debris. The remainder of that layer may be backfilled mechanically. Break up large clumps as much as possible. Backfill the trenches and compact according to specifications appropriate to the locale.

Penetrating the Building Envelope. The subgrade portion of the mechanical room walls or floor is at common penetration locations. Separate the supply and discharge lines by at least 2 feet. Penetrations through floors must be cemented in, and those through walls must be sealed using wall sleeves. The outside of the wall penetration must be caulked with suitable material and sealed with cold-applied asphalt bitumen waterproofing. Allow this to cure prior to backfilling. Pipe must be protected against differential settlement at the wall.

Connection to the Plate Frame Heat Exchanger. When multiple wells with submersible pumps are employed, all of the supply pipes are connected through a header, installed in the mechanical room, to the entering well-water port of the plate frame heat exchanger. The well water exiting the heat exchanger is also headered when multiple disposal wells are used. Isolation valves on each header leg allow individual wells to be switched in and out of production and also permit the use of one well for either production or disposal purposes. It is important to allow sufficient room for removal of the movable end plate and heat transfer plates of the plate heat exchanger.

Other Considerations. The following are additional considerations at the time of groundwater system specification preparation or installation:

1. The return well(s) shall be constructed to ensure no air entrainment, which in the presence of iron bacteria can cause clogging of the well(s).

2. Avoid aeration of recharge water prior to injection. The presence of oxygen forms oxides of iron-scale on the return well screen and can also cause clogging. The discharge connection to the return well should be several feet below the lowest anticipated static water level and the remainder of the groundwater delivery system should be airtight and not cause induced fluid turbulence in the return well.

3. Where return water is discharged to a surface water body or storm sewer, it shall be done with the permission of the local authority in accordance with local codes.

4. Specify plastics in the piping network and appropriate materials in pump housings and in the plate frame heat exchanger.

5. Specify flow rate, temperature, and pressure instrumentation both upstream and downstream of the heat exchanger.

On-site testing, verification, and reporting should be undertaken as described in the following:

1. Pipes of the correct dimension and materials have been employed.

2. The trenches and final grading are in accordance with design.

3. Piping connections have been correctly executed and are leak free.

C.3 HEAT PUMP AND INTERNAL DISTRIBUTION SYSTEM

Piping System. Following completion of the building piping system and installation of the mechanical room piping, it remains to thoroughly clean the system and remove any air. This is normally the responsibility of the mechanical contractor but should be observed by the A/E as commissioning agent. The ground-source system piping shall remain isolated from the building and mechanical room piping during this procedure. This part of the installation involves:

1. Filling the building piping system with water using the water fill line in the mechanical room. Heat pump heat exchangers and other equipment intended to be connected to the loop are to be bypassed during this operation.
2. Keep running water through the system until the debris, dirt, etc., are cleared from the piping. Large diameter piping may have to be flushed separately to ensure high enough velocities to remove debris.
3. Connect the heat pumps and other conditioning equipment to the loop and once again fill the system with clean water. Open all air vents at the top of the system; open all isolation valves on runouts at heat pumps, and open the isolation valves on the water fill line in the mechanical room.

   Note: Building water pressure must be sufficient to overcome the hydrostatic head between the fill line and the high point of the system.
4. Turn on the building water and close the manual air vents when the system is filled and water begins to flow from air vents.
5. With the fill line still open, flush the system again using building water at fill line and open drains at the lowest elevations. Runout and riser isolation valves must be opened and closed sequentially to ensure that each major parallel path is flushed individually and finally as a full system.
6. The process should then be repeated using the building water-loop pumps.
7. Where metal piping is used in the water-loop, chemical cleaning is recommended prior to final charging.
8. Add an aqueous solution of trisodium phosphate (one pound per gallon of water in the system) and raise the system temperature to 95°F and circulate for two hours. Use a temporary heater in the circuit, if need be.
9. Any leaks shall be repaired. No stop leak compounds shall be added to stop piping system leaks.
10. At the end of the two hours, drain, flush, and refill with fresh water and test to ensure that the water is left slightly alkaline (pH of 7.5). If still on the acid side (less than pH of 7.5) repeat the chemical cleaning.

If the system has a ground heat exchanger or surface water heat exchanger, the antifreeze was added earlier to the heat exchanger and isolated from the indoor system. The remaining procedure is as follows:

1. Open isolation valves so that the ground heat exchanger/surface water heat exchanger and indoor water-loop have one continuous flow path. Operate the system pump to completely mix the fluid throughout the system.
2. Verify that the freeze protection is within the requirements of the design.
3. Verify that the chemical inhibitor characteristic and concentration are as called for by the design.
4. The latter two tests should be undertaken by a third party, independent contractor.

A flow balancing report shall be prepared by a balancing contractor. The work involves the following:
1. Flow shall be balanced through all parallel paths in the system to the requirements of the design.
2. Set the system in operation with one of the circulation pumps.
3. Set the total system flow across the pump in the central pumping station using the differential pressure measured across the orifice.
4. For multi-story buildings adjust the flow to the design for each floor or runout using the balancing valve or circuit setter in the runout return leg at each floor.
5. Balance flow to each heat pump or parallel device on each runout using the pressure drop readings across each heat pump or device and comparing with manufacturer's data. Adjust the pressure drop as needed using the balancing or ball valve on the return line of each heat pump.
6. The flow balancing report shall show the main riser or branch GPMs and the measured flows across each heat pump or other device on the loop.